BOTANICAL NOMENCLATURE

BOTANICAL NOMENCLATURE

A. ROSELINE

Professor and Head
Department of Botany
Holy Cross College (Autonomous)
Trichy, Tamil Nadu

MJP Publishers

Chennai New Delhi Tirunelveli

ISBN 978-81-8094-197-9
MJP Publishers

Printed and bound in India
New No. 5 Muthu Kalathy Street,
Triplicane,
Chennai 600 005
MJP 170

Publisher : J.C. Pillai

PREFACE

The study of Botany cannot be fulfilled without learning the system of botanical nomenclature, the backbone on which edifice of entire botany is built. In recent years teaching of plant systematics has become a challenge to the teachers, and also teaching of taxonomy in the class room is incomplete if botanical nomenclature has not been taught properly to the students to identify and name the plants correctly.

Although sufficient literature is available, no effort has been made to provide an organised manner, so that the readers would find it very easy to comprehend the basic knowledge of the Code of botanical nomenclature. Moreover, there is an urgent need for the book on relevant aspects of nomenclature to be published for reference. To overcome the hurdles of such situation, the present book is comprehensively prepared to cater to the need of students and teachers of botanical science who have limited exposure to nomenclature of plants. The present attempt is a compilation of all my personal reference exracted through long journey of teaching plant taxonomy.

I wish to record my sincere gratitude to my principal DR.(Sr.) Jeusin Francis for her constant encouragement and moral concern rendered to complete the work successfully. My heartfelt thanks are due to my own sisters DR.(Sr.) John Marie and DR.(Sr.) Metti for their prayerful support, and to my teachers, DR.(Mrs.) P. Francisca, Associate professor of my department and other colleagues, supportive staff of my department and also other welwishers for initiating me to write the book on plant nomenclature.

It is my bound duty to appreciate whole heartedly and acknowledge my dear friend DR.(Miss.) D.Radha, Associate Professor of Economics department of my college for her timely help extended towards the correction of manuscript and also for her untiring inspiration which boosted me to proceed in a very smoothway to complete the book.

I also wish to express my indebtedness to my family members Mr. S.Xavier my husband and my son X.Infant Reni B.E., for the personal sacrifice made and also for offering their full co- operation in various capacities during the preparation of this text.

Finally I thank in a very special way the publishers of the book for being very patient with me and doing an excellent job.

A. Roseline

CONTENTS

1

INTRODUCTION

Keywords Definition, botanical nomenclature, common names vs. scientific names, biological language, objectives of code, operative principles.

Definition The method of naming plants on a scientific basis is known as botanical nomenclature.

Any object once identified on this earth has to be given a name to communicate and to differentiate it. This art of giving names to objects is referred to as nomenclature. The word nomenclature is derived from two Latin words, 'nomen'- means 'name' and 'clare'- means 'call ', i.e., the scientific art of naming an object. Likewise the plants that are known to us should also be given with names for identification, communication and also to differentiate and describe it. The ultimate function of botanical nomenclature is to provide a correct name to each taxon.

Assigning of names to plants has a very long history, which dates back to the dawn of human civilization.

Early attempt In the beginning, the names were given to plants by people in a particular area or locality. These names are called vernacular or local names, which differed from language to language and even from dialect to dialect. Therefore, the common names were restricted to a particular area or a population, but with the advancement of knowledge more and more common names were accumulated. As a result, a particular plant may have several names in different localities of a particular country. This has ultimately created a lot of confusion in understanding the names of plants.

Thus, in the beginning almost all the useful plants were given only with local or common names which varied from place to place. Even in Ayurveda, the ancient Indian literature on medicine plants, for e.g., the Mango (*Mangifera indica*) is known by over 50 different names, all in the Sanskrit language. Hence, common names obviously have a limited usage and also proved to be misleading and inadequate for worldwide use.

Therefore, the botanists felt the need of having scientific way of proposing names for plants so that, it could be accepted universally.

Need for scientific names The universally proposed scientific names for the plants has the following advantages over the common names (local names), which created many problems. They are as follows:

- Common names vary from place to place.
- It is restricted to the people of one particular linguistic group. In multilingual country like India, the species *Ipomoea batatas* is called sweet potato in English, and in Hindi and in Punjabi it is known by its common name Shakarkand; in Bengali, Lal Aalu; in Telegu, Chelaganda and in Tamil it is called charkaraivalli Kizhanghu.
- Common names usually do not provide proper information of the generic or family relationship. For e.g., Roses belong to the genus Rosa, whereas Woodrose is a member of the genus *Ipomoea* and Primrose belongs

to the genus *Primula*. The three genera in turn belong to three different families such as *Rosaceae*, *Convolvulaceae*, and *Primulaceae* respectively.

- Sometimes different plants may have the same common name. For e.g., the name Ironweed is given to the plant *Sida* of the family *Malvaceae* and also for the plant *Vernonia* of the family *Asteraceae*. Similarly Cockscomb is a common name for *Celosia cristata* but is also applied to *Rhinanthus minor* or to a Seaweed *Plocamium coccineum*.
- Widely distributed plants and the plants with high economic value have a larger number of common names. For e.g., Pancy (*Viola tricolour*) an ornamental plant commonly grown in most parts of the European and American countries has more than 150 names , in addition to 50 common names in English. Similarly *Plantago* with more than 50 common names and *Verbascum* has over 140 vernacular names.
- A well known plant may have different names in different linguistic area and even in different parts of the same area. For e.g., *Chrysanthemum leucanthemum* is commonly called Daisy, White or Ox-eye daisy, Shasta daisy or White weed.
- The plants that are not used by the local people were not given with names.

Hence the common names are neither methodical nor universal often proved to be misleading and inadequate for wider use throughout the whole world.

In order to overcome the hurdles raised by the usage of common names, the scientists felt the need of having scientific way of proposing names for plants. The significance of scientific names are as follows:

i. The scientific names are universal and are given to all known plants on earth irrespective of their use.

ii. The scientific names usually give information about the plants of families.

iii. Each plant has one correct name except in case of synonyms for certain species. Therefore plant should be named in such a way so that it can be accepted universally.

While giving scientific names to plants, the first problem which the scientists faced, was to select a suitable language which can give a clearcut meaning and which would not change. The botanists of the world unanimously agreed to have Latin as the language for biological nomenclature for the following reasons:

i. Firstly Latin being a dead language, no longer spoken in any country or nation as a native tongue by any people, hence it avoids any national bias.

ii. Secondly most of the European languages are derived from Latin.

iii. Thirdly in the past Latin was a widely used language during the Middle Ages throughout the European countries, where a lot of botanical literature has been written in Latin. Also, since Europe, particulary Greece, has dominated the whole world in the field of science in general and Botany in particular during 1600 B.C. to 1850 A.D, where the science of botany originated and first developed. In addition by 1780 about 80 per cent of the plants were already named in Latin.

iv. Fourthly latin is precise, concise and therefore it is specific i.e. it gives the exact meaning and genetic relationships of generic names and is unambiguous in the grammatical sense of the word as is commonly obvious e.g., when translated as album- neuter, alba-feminine or albus-masculine.

v. Lastly latin is pertinent to the needs of descriptive phrases of natural science because it employs Roman alphabet, which fits well in the text of the most languages.

Therefore Latin forms the language of biological nomenclature.

The present system of nomenclature is the result of a historic series of changes that gradually became formalised. Today all plants have a Latinized names regardless of their origin.

History of Botanical nomenclature Naming of plants involves principles governed by rules developed and adopted by International Botanical Congress. The rules formulated are formally listed in the International Code of Botanical Nomenclature (ICBN), which is published by the International Association of Plant Taxonomy (IAPT) and are often referred to as a Code by Stafleu (1972). The International Code of Nomenclature for Cultivated Plants (ICNCP) regulates the naming of cultivated plants, which is largely based on the ICBN with a few additional provisions.

The ultimate goal of this system is to provide one correct name for each taxon. Much effort has been devoted in establishing procedures for naming taxa and for changing names that were incorrectly assigned. Nomenclature forms the basic groundwork, on which the whole edifice of both taxonomy and the study of entire botany are built and it forms an important function of plant nomenclature.

Naming of plants has a very long history from the days of Theophrastus (350 B.C.). In the beginning names given to plants were long descriptive phrases. For e.g., *Lychnis alpina linifolia multiflora perampla radice*, i.e., high mountain lychnis with large root and leaves like flax and with several flowers.

This system of giving several words to a plant was named as polynomial system of nomenclature and the names are called **polynomials**.

This system was in practice up to the middle of 18th century. Although these names were meaningful, it became difficult to remember such long names, when the number of plant species increased. Therefore, the polynomial system did not continue further.

Later **Gaspard** (Caspar) Bauhin (1560-1624) came up with the novel idea of giving only two words for every plant and the first name of the plant is the genus followed by the second name, the specific epithet. This system of naming of plants with only two words is called **binomial nomenclature**. But at that time this system was not realised or recognised by many taxonomists. However, at least 100 years later, Linnaeus (1707-1778) a Swedish botanist gave life to binomial nomenclature. According to Linnaeus, each plant was given with two names. The first name is the generic which actually refers to the plant, and the second name is the specific epithet which refers to the species of that plant.

Therefore, the method of naming of plants with two names that is generic and specific epithet is called **binary** or **binomial nomenclature** and the name is called **binomial or binary name.** This system was introduced by Carolus Linnaeus in 1753 in his *Species Plantarum*. For e.g., the binomial *Mangifera indica* is given to the plant Mango, where the first name, the genus *Mangifera* indicates the plant mango and the specific name *indica* denotes the particular species of *Mangifera.*

Later in 1974 Heywood proposed another system of naming plants called uninomial system of nomenclature. According to uninomial system each plant was given with only one name. It also had certain advantages for different purposes.

This system was proposed to overcome certain difficulties faced by adopting binomial system, especially, when there is a shift from one genus to another genus carrying the specific epithet of the previous name, which ultimately resulted in duplication of names in the same taxon.

However, the binomial system of nomenclature has been widely accepted and adopted even today because of its excellent merit. To overcome the discrepancies in nomenclature due to the absence of general rules, Linnaeus proposed various Laws and Aphorisms for plant nomenclature in his *Fundamenta Botanical* (1736 and *Critica Botanica* (1737). Later, in *Philosophia Botanica* (1751) he amplified the views of aphorisms and principles in detail, thereby establishing a real beginning of a sound system of plant nomenclature. Earlier there was no generally accepted rules for naming of plants except the Linnaean principles, and in different countries every botanist followed his own way in giving names to plants.

According to Linnaean principles,

i. No two genera could have the same name and no two species within a genus could have the same specific epithet.

ii. If a genus was divided into two or more genera, the original generic name of the genus must be retained for one among them.

iii. Priority of publication was recognised in principle.

iv. He employed binomial system to avoid assigning different names to a species or identical names to different species.

v. When a variety was recognised, it must be associated nomenclaturally with that species as a variant.

Since plants new to science have increased so rapidly, confusion in nomenclature existed. There was no index of plant names from which botanists of different countries could learn the name of plants that had been used.Later in 1813 Augustin P.de Candolle published *'Theorie elementaire de la botanique'* containing instructions on various nomenclatural procedures mostly followed by Linnaeus. Steudel published his *'Nomenclator botanicus'* (1821), indexing the names of flowering plants with their synonyms. This was useful for checking names and synonyms. It was the forerunner of *'Index Kewensis'* (Lists of plant names published by Kew garden). But, the increased number of plants known to European botanists in the 18th century required the development of order and stability in plant nomenclature. The acceptance of priority of name was recognised as the cornerstone of nomenclature, ensuring that each plant had a unique name.

Hence, the International Botanical Congress (1866) appointed the great French botanist Alphonse de Candolle to frame the rules of nomenclature for plant taxonomists. His report was accepted at the first congress held in Paris (1867). This first congress, which held at Paris,was known as *'Paris Code'* or the *'Candollean Code'* and Alphonse de Candolle, the son of Augustin P. de Candolle, headed this first International Botanical Congress (IBC). Botanists from different countries met and adopted a set of rules for nomenclature of plants. However, Alphonse de Candolle had proposed most of the rules. Although the rules framed at the beginning were excellent, later the botanists felt the need to modify the rules in late 1800s. Hence, Botanical Congresses were held in different countries periodically once in 5 years to resolve nomenclatural problems and to standardise on a worldwide basis with accepted rules by the members of the IBC, with additions, deletions and modifications. The rules adopted at the congresses are published in the form of ICBN and published in the Journal of '**Taxon**'.

Purpose of giving names The plants are given names for two main reasons:

1. As an aid to communication and
2. To indicate their relationships.

In practise nomenclature is often combined with identification. This is a fundamental principle of menclature and it is the most important criterion by which the efficiency of any system of nomenclature can be judged. It implies that names must be unambiguous and universal and this is because the scientific names of plants must avoid any possibility of such confusion and it should be the same throughout the world.

The Aims of the Code The Code tries to ensure that with any given circumscription, position and rank, each taxon can have only one correct name by which it may be properly known. Naming the plants acceptable to all the botanists in the world is the system of international importance and it is an important function of taxonomy adopted by the ICBN. The Code also tries to avoid or reject certain names and retain some of the names to achieve these aims. In addition, the Code also lays down certain provisions, which must be followed in giving names to taxa and in the use of names. These provisions are based on the number of what may be called the operative principles of which, the most important are

1. Effective and Valid publication
2. Typification and
3. Principles of priority.

According to ICBN,

1. no two genera could have the identical generic name.
2. no two species within a given genus could have the same species name.

3. when a genus is divided into two or more genera, the original name must be retained for any one of them which has the type and
4. priority of publication was given prime importance.

2

POLYNOMIAL SYSTEM OF NOMENCLATURE

Keywords Definition, polynomials, polynomial system, trivial names, examples.

In the early attempts, scientific plant names were originally given in the form of short descriptive terms or additional adjectives to distinguish between the species, which ultimately gave rise to multiple terms called polynomials.

For e.g., *Lychnis alpina linifolia multiflora perampla radice* i.e., high mountain lychnis with large root and leaves like flax and with several flowers. This system of nomenclature was known as **polynomial system of nomenclature** and the plant named as **polynomial**. Many such names have their origin from Latin, a language known to learned people of Europe where science of botany had its origin.

Polynomials have been used by many taxonomists including Linnaeus (1753).The polynomial system was used in the naming of plants until the first half of 18th century. Polynomial system means, a plant name composed of several words in a series, constituting more or less a description or diagnosis of the plant. Linnaeus used polynomials for all the plants in his Species Plantarum. It was only for the sake of convenience that he used additional trivial names for these polynomials, which later on helped other taxonomists to use binomial for plants.

Thus the species *Sida cordifolia* Linn.(1753) as it is known today, was named as '*Althaea maderaspathana subrotundo folio molli et hirsuto multipilis* ' by Plukenet (1692). Similarly *Strychnos nux-vomica* L. was then known as *Strychnos foliis ovatis caule enerme* and *S. colubrina* L. as *Strychnos foliis ovatis, acutis cirrhis simplicibus*.

Similar examples can also be seen in many earlier works of Linnaeus. In 1738 Linnaeus named a species as '*Plantago foliis ovatis glabris* '. This phrase had to be amended, when the number of species increased and these characters also became insufficient for unambiguous differentiation of the taxon. So, in 1753, the above mentioned species by Linnaeus became '*Plantago foliis ovatis glabris, nudo scapo tereti, spica flosculis imbricatuis*', which was later on named as *Plantago major*. These polynomials served as labels for the taxa and at the same time for diagnosis of the species. These two functions ultimately proved to be incompatible and consequently Linnaeus separated them.

Although, these names were meaningful it became very difficult to remember, and also indexing such long plant names when, the number of plants being increased. As a result, this polynomial system could not be continued further. Gaspard (or Caspar) Bauhin (1560-1624) in his monumental work '*Pinax*'(1623) devised a novel idea of giving only two names to every plant, called binomial system, by differentiating the generic name and specific epithet of the taxon. Unfortunately, his system was not consistent in its use at that time. After that, Carolus Linnaeus rightly adopted the binomial system of naming plants in true sense. Accordingly, each taxon was given with only two names such as a generic name followed by a specific epithet.

Thus, the system of polynomial was replaced by binomial system, which was introduced by Carolus Linnaeus in the first edition of his *Species Plantarum* published in the year 1753.

3

BINOMIAL SYSTEM OF NOMENCLATURE

Keywords Definition, binary name, guiding principles, Linnaeus monumental work, polynomial vs. binomials, binomial system, infraspecific taxa, examples, hybrids, cultivars, significance.

The system of naming plants by two words consisting of a generic name and a specific epithet is called **binomial system of nomenclature**.

The binomial nomenclature was truly adopted and employed firmly and consistently by a Swedish botanist Carolus Linnaeus in the first edition of his *Species Plantarum* in 1753, although this system was infact first suggested by Caspar Bauhin (1560-1624) at least hundred years earlier in his publication *Pinax (*1623). The work of *Species Plantarum* of Linnaeus described about 6000 species in 1000 genera and is considered the most important monumental work in the history of botanical world today.

The two words in combination constitute a plant name i.e. the generic and a specific epithet is called as **binary name** or **binomial name.** The binomial system was universally accepted by all botanists and is still in the usage. Thus, each taxon therefore, has two parts, the first name is the genus, and the second is the specific epithet (trivial name) in which, the first name is given to the plant of one type known as generic name and the second name is called the specific epithet, which denotes the plant of similar kind with slight differences.

For example, the botanical name of Rose is *Rosa indica* and in this, the first word *Rosa* designates the genus of the Rose plant and the second word *indica* denotes a particular species of that genus. The name of the person who first described the binomial must be suffixed to it. E.g. *Rosa indica* Linn. The two words in combination constitute a binomial and is derived from Latin language.

However, it should be noted that the generic name *Rosa* alone conveys the meaning of the Rose plant but the specific epithets alone do not convey any sense unless combined with generic names. All the botanical names are written in Latin language. There is only one botanical name for each plant known as a legitimate name.

According to the International Rules of Botanical Nomenclature, the generic name is always a singular noun and written with a capital initial letter and the specific name is an adjective of the same gender and it starts with a small initial (lower case) letter. Thus in the binomial system, a botanical name of a species is a combination of a generic name and a specific epithet. For example, the botanical name of Rose is *Rosa indica*. The scientific botanical names are always printed in italics and if typed or hand written should be always underlined or boldface.

The generic names may have different origin and it may be named in honour of a botanist, e.g., *Linnaea* for Linnaeus, *Caesalpinia* for Caesalpino, *Hookera* for Hooker, etc. It may be from mythological or poetical origin e.g., *Nymphaea* (water lilies) denotes the lovely water-nymphs, or it may be named after the characteristic features of plants e.g., *Hygrophila*, denotes the marshy habit, liver- like leaves of *Hepatica*, *Xanthoxylum* indicates yellow colour of the wood, *Liriodendron* (lily tree) from the shape of the flowers of the Tulip tree, or based on a place such as *Arabis* for Arabia. Sometimes it is of aboriginal in origin (language) i.e., taken directly from a language other than Latin. For example *Tsuga* from Japanese, *Nelumbo* from Ceylonese, *Betula* from Birch, *Vanda* from Sanskrit, *Ginkgo* a Chinese origin and so on. The species are regarded as morphological definable units which can breed true among its individuals. These specific names indicate the kind of plant and enable the members of a genus to distinguish from each other.

The specific epithet is an adjective indicating a distinguishing characteristic of the species for example *alba* refers to white colour, or referring to a locality where the species was first discovered e.g., *indica* from India or in honour of a renowned botanist like *Santapauii* after Rev. Fr. H. Santapau, or something that indicates the habitat of the plant e.g., *aquatica* (in water).

The specific epithet is always written with a small initial letter unlike the generic name. However, if the specific name is derived from a person's name, the initial letter of the epithet may be capital letter e.g., *Pinus Roxburghii* (from Roxburgh).

The specific epithets are referred as 'trivial name' by Linnaeus and he adopted this system of trivial names unintentionally, however it was found very useful to the later workers to adopt a generic name and followed by a trivial name (specific epithet). If an epithet or name has, two or more words must be united or hyphened.

For e.g., *Trigonella foenum- graecum.*

In the case of an infraspecific taxon like subsp, variety, subvariety, form and sub form, the epithet should be connected by the term denoting its rank and these ranks are also in Latin and are adjectives (Art.24).

E.g., 1. *Crotalaria willdenowia* subsp. *glabrifolia.*

2. *Rosa gallica* var. *gallica.*

3. *Prunus cornuta* form. *villosa.*

The generic names came first, for a century and a half; species were designated by more or less descriptive phrases, which could be changed by any author for a purpose.

The names of cultivated plants are governed by the International Code of Nomenclature for Cultivated Plants (ICNCP). Cultivar names must not be written in italics but they can start with a capital initial letter and the names may be used after generic, specific or common names. The rank of cultivar is abbreviated as cv. for cultivated varieties.

For example,

1. Camelia japonica cv.Purple Dawn for cultivated varieties.
2. Citrullus cv. Crimson Sweet.

Or

Citrullus lanatus cv. Crimson Sweet.

Hybrids are usually designated by formulae, which indicate their hybrid nature and putative parents as given below:

1. The names of both the parents are connected by a multiplication sign.

   e.g., i. *Verbascum lychnites* × *V. nigrum*

   ii. *Salix aurita* × *S. caprea*

2. If a new epithet is provided for the hybrid, the generic name and the epithet are connected by a multiplication sign.

   e.g., i. *Verbascum* × *schiedianum*,

   ii. *Salix* × *caprea*

The names of intergeneric hybrids are framed by a combination of both the generic names preceded by a multiplication sign as cited below:

e.g., i. *Triticum* × *Secale* = × *Triticale*,

ii. *Agrostis* × *Polypogon* = × *Agropogon*,

iii. *Agropyron* × *Hordeum* = × *Agrohordeum*.

The provisions for hybrids (Appendix I) were thoroughly overhauled at Sydney. However, there were no significant changes made except for the adoption of the word *nothotaxon*, to refer to a taxon, at any rank, of hybrid origin. Thus, now we have *nothogenera*, *nothospecies*, etc., for hybrids. The grafts between two species are indicated by a plus sign (+) in front of the name of graft hybrid (+ *Crataegomespilus dardari*), or between two grafted species, for e.g., *Rosa webbiana* + *R. floribunda*. For cultivated plants there is a separate International Code for Nomenclature of Cultivated Plants.

Thus, the modern binomial system was born and this system superseded the earlier difficult polynomial system. Later the application of binomial to the vegetable kingdom as a whole was very efficiently given and constantly practised by Carolus Linnaeus in his *Species Plantarum* (1753). The method of giving only two names to a plant was proved to be so convenient that it was universally accepted by all botanists of the world and is in practise up to the present day.

Although many of the names of plants proposed by Linnaeus were from earlier authors, for practical purpose,

Species Plantarum (1753) of Linnaeus' work acquired so much importance in plant taxonomy over the years. On 1st May, 1753 *Species Plantarum* was published and is treated as the starting date of botanical nomenclature for most groups of plants including the Angiosperms. The elementary rule on which he based his system of nomenclature are dealt in his *Fundamenta Botanica* (1736), and *Critica Botanica* (1737) and he further amplified the rules in his *Philosophia Botanica* (1751). These rules of botanical nomenclature later served as guidelines to other botanists of the world.

Advantages of Botanical Nomenclature

1. Binary names are precise than ordinary names and more definite.
2. These names being in Latin language, people of all languages universally accepted them.
3. These names are more comprehensible, very easy to remember and indexing.
4. These names indicate the genetic relationship and descent of individual plant.
5. They are more independent and are usually descriptive.

4

UNINOMIAL SYSTEM OF NOMENCLATURE

Keywords Objectives, definition, uninomials, uninomial system, reference number, uninomial vs. binomials,examples.

The botanists while following the binomial system of nomenclature as proposed by Carolus Linnaeus greatly felt certain practical difficulties. Firstly, the rules for naming plants were slightly complicated and secondly, it was very difficult to shift one individual species from one genus to another genus, carrying the specific epithet of the previous name. This often resulted in duplication of names in the same taxon, which is contrary to the current system of nomenclature. According to the rules, due to change in position of the individual in taxonomic hierarchy, both generic and specific epithets are to be changed. The change in name and duplication of names is a serious problem, which is faced by the taxonomists while changing the systematic position of any plant.

The present day taxonomy is progressing with the added information from various branches of botany such as anatomy, embryology, palynology, cytology, genetics, taximetrics, biochemistry etc., also helped in the establishment of species concept. As a result, closely related plant species are placed wide apart and this necessitated the shifting of morphologically allied species quite apart.

When change occurred in the rank of taxa, this sometimes resulted in the formation of tautonyms, which are inadmissible to the rules of botanical nomenclature. Therefore, in order to simplify the system of naming plants and to overcome the above mentioned problems, Heywood (1974) suggested that the two names, namely generic and specific names, may be united together into a single unit or hyphenated to form a single unit, so that in case of shifting of one species to another taxon, there is no necessity needed to introduce a change in generic name or specific epithets. For e.g., when *Ranunculus ficaria* is shifted to genus *Ficaria,* it need not be *ficaria* but as *Ranunculus ficaria,* regardless whether it belongs to *Ficaria, Ranunculus* or any other genus.

Another alternative method was also suggested for each taxonomic group, two numbers may be given, which would remain unchanged throughout, irrespective of any change in its position and relationship. This number is termed as *reference number*, which would carry the descriptions along with them. The method of giving number is more convenient for taxonomists, since the knowledge about plants is greatly expanding and it would be very difficult to remember all the names and characters regarding plants. The coding of reference numbers with description of characters was also suggested. Reference Numbers would anatomically give the knowledge about the plants, with the help of Code. Thus, the system of giving one name to the individual plant is termed as **uninomial system of nomenclature**, and it has certain advantages for different purposes. The uninomial system was proposed by Heywood in 1974.

In early days, Theophrastus (370-285B.C.) and later Pliny (32-75A.D.) and Roman workers used Latin names for plants as *Rosa* for rose, *Pyrus* for pear, *Viola* for pansy, *Morus* for mulberry and *Ficus* for fig, and these names were called **mononomials or uninomials**. Although these names were used for a long time, difficulty arose when different species of the same plant could not be expressed, and also these uninomial names did not refer to the exact species in the sense as recognised in Linnaean and post-Linnaean periods. Therefore, the uninomial system was replaced by binomial system of Linnaeus to differentiate the different species of the plants.

5

HISTORY AND DEVELOPMENT OF THE CODE OF BOTANICAL NOMENCLATURE

Keywords Introduction, Linnaeus' principles, Index Kewensis, code of conduct, botanical congress, code, Kew rule, valid name, principle of priority, latin name, type specimen, recommendations, taxonomic group, legitimate & illegitimate names, effective and valid publications, typification, nomina conservanda, appendices, hybrid names.

Introduction The foundations for the rules of present day's nomenclature had their origin in Linnaeus' *Critica Botanica* (1737), which was an elaboration of his own *Funtamenta Botanica* (1736). This was further amplified in his '*Philosophia Botanica*' (1751) and established the real beginning of a sound nomenclature for plants by proposing elementary principles. There were no rules except for Linnaeus' Principles of nomenclature. These rules of botanical nomenclature served as a guide to the botanists.

Another significant work on plant nomenclature was the Law of Botanical nomenclature framed by Augustin de

Candolle in his '*Theorie elementaire de la botanique*'(1813), which provides explicit instructions on matters of organography and procedures of botanical nomenclature. Which include a detailed rules on plant nomenclature, which was mainly taken from guiding rules of Linnaeus. It also pointed out that the fallacies of practices and proposals of other contemporary botanists. This was probably the first complete and detailed set of rules on plant nomenclature.

The first important post- Linnaean index was E.G.Steudel's *Nomenclator Botanicus* (1821) which comprised a list of Latin names of all flowering plants known at that time along with their synonyms. The second edition appeared in 1840, and it was a forerunner of *Index Kewensis* published half a century later by Kew garden.

However, the increase in number of newly explored plants caused the concern over (procedure for) naming of these species, and Linnaeus underestimated the diversity among plants. The post-Linnaean workers also estimated about three lakh plants as against 10,000 presumed by Linnaeus. So the rules proposed by Linnaeus ran into trouble as these were not sufficient enough to cover the plant nomenclature. The same species was described under different names from different parts of the world by different authors, which also ultimately resulted in about eight and a half lakh names as recorded in the *Index Kewensis* for the period 1753-1986; whereas the actual number of species reported to be about ⅓ as recorded in the *Index Kewensis*. This increase in numbers led to confusion in nomenclature.

More over in different countries, nomenclature followed different patterns and caused personal and national jealousies. National and personal jealousies further complicated the situation and thus the need for internationally accepted rules for nomenclature was very much felt. Thus over a period of time, the need for an international accord became increasingly apparent.

At last, the International Botanical Congress(IBC) (1866) appointed the great French botanist Alphonse de Candolle, the son of A. P. de Candolle to draw up 'a *code of conduct*' (rules of nomenclature) for plant taxonomists in nomenclatural matters and it was accepted at the first Botanical Congress held in Paris (1867), which is known as '*Paris Code*'. It is also otherwise called '*Candollean Code*'. In Paris Code, the date of publication of *Species plantarum* (1753) of Linnaeus was made as the starting point for plant nomenclature and rules of priority (Art.15) were made fundamental.

Not satisfied with the Paris Code, American botanists conducted a separate meeting at Rochester in 1892 to frame the rules and referred to as Rochester Code (1892), which introduced the concept of types, and also insisted the strict application of rules of priority even if the name was a tautonym.

The subsequent meeting of the IBC was held once in five year at different countries. Therefore the Paris Code was replaced by the Vienna Code (1905), which also accepted the *Species plantarum* (1753) of Linnaeus as the starting point for vascular plants. Tautonym was not accepted and Latin diagnosis was made essential for new species. In addition, a list of conserved generic names (Nomina generica conservanda) was approved. With the dissatisfaction of the Vienna Code, the adherents of the Rochester Code adopted the American Code (1907) which did not accept the list of conserved names and also the requirement for Latin diagnosis and in the year 1910 the Brussels Code was adopted by IBC.

But a general agreement regarding the ternationally accepted rules of plant nomenclature was attained in 1930 at the 5th IBC held at Cambridge for the first time in the botanical history, a Code of nomenclature ie., International Code of Botanical Nomenclature (ICBN) which was accepted internationally came into practise. Therefore this Code is called the International Code of Botanical Nomenclature. The modification or amendment, as suggested by IBCs at subsequent meetings have regularly been incorporated in the Code. The details of the various Code adopted during different IBCs are described below:

1. **Paris Code (1867)** The first organized Botanical Congress was held at Paris in August 1867, under the headship of A P de Candolle for standardisation and legislation of proper nomenclature practice. A copy of the law of nomenclature was called as *Lois de la Nomenclature Botanique* prepared by Alphonse de Candolle was circulated in advance to about 150 European and American botanists who were invited to attend the congress to present a new set of rules at the assembly. The law of botanical nomenclature was mainly based on his father's (Augustin de Candolle) *Theorie elementaire de la Botanique* (1813), which in turn was based on various aphorisms and pronouncements of Linnaeus. The Law of botanical nomenclature was discussed and with slight revision it was resolved and adopted by the assembly, shall be recommended as the best guide for nomenclature of plant kingdom. The law was adopted at the French capitaland the rules are called '*Paris Code*' or *de Candolle's Rules* as it was proposed by Alphonse de Candolle.

According to the Paris Code, the starting point for all plant nomenclature was fixed with Linnaeus. However, no date or any work was specified. The Paris Code consisted of three chapters having sixty eight articles. The rules, agreed along with other matters in the Paris Code, include the following:

- The rule of priority was considered as the basic with no provisions of exception.

- Emphasis was laid on author citation. It was suggested that the person who publishes the account for any plant for the first time and gives a new name to the genus should be considered as the author of that scientific name.
- A plant can have only one valid name.
- A species is not named unless its specific name is assigned to a genus.
- The name of the plant must be accompanied by a description or other means of identification, with date.
- Two different species cannot bear the same name.
- It also emphasizes the discredibility of Latin name of plants.
- When a species is transferred from one genus to another genus, the first specific epithet is retained.
- Many abbreviations of authors were also recognized like L or Linn. for Linnaeus, Hook.f. for J.D.Hooker, D.C. for de Candolle, R.Br. for Robert Brown etc.

Kew Rule Later on, many American and British (English) botanists who were not satisfied with the Paris Code, deviated from it and practised a law of their own called *Kew Rule,* which was considered as an unwritten law. According to Kew Rule, if a species is transferred to another genus, the original specific epithet need not be transferred to the new genus, instead the author was free to use a new epithet in the new combination. As days go on, the Kew Rule led to much controversy between British botanists who supported the rule and continental and the neo- Americans who opposed the rule. However, the Kew Rule was followed in the preparation and compilation of the first two volumes of the Index Kewensis.

2. **Rochester Code (1892)** A batch of the botanists from the United States headed by N.L.Britton of New York Botanical Garden proposed a set of rules to govern plant nomenclature. These rules were presented for discussion at the meeting of the Botanical Club of American Association for Advancement of Science held in Rochester, New York in 1892. These were based on modification of the Paris code. These modified rules are commonly known as Rochester Code.

Some of the new recommendations added in the Rochester Code are as follows:

- Establishment of the type concept to ascertain the correct application of the names.
- Strict application to the principle of priority.
- Acceptance of alternate or all binomials resulting from application of the principles of priority, even if the specific epithet repeats the generic name (referred to as a tautonym) e.g,. *Malus malus.*
- When a species is transferred from one genus to another then the specific epithet is carried by an individual and the generic name can be subjected to a change.
- Interpretation of priority to the precedence of a name in a publication in addition to the date of publication.
- Publication of each new species must be based upon a designated herbarium specimen to be known as the 'Type Specimen' (nomenclatural type), and it should be deposited and preserved in any recognized herbarium.

Weatherby (1949) stated that, this Code was brief, and clear, and it seemed to resolve all the difficulties. Rochester Code was followed by the majority of the American taxonomists, but not by the Europeans. Therefore, these rules created two opposing schools of thought among the American botanists, one school was headed by N.L.Britton and his students, and the other by the students of Asa Gray Others having more conservative philosophies followed the Kew Rule. Keeping this in view, need for the revision of rules was felt very much, and in 1900 a commission was made with 40 members headed by John Briquest of Geneva. He was asked to draw up a revised Code, which form the Vienna Code 1906.

3. **Vienna Code (1906)** The Third International Congress conducted in Vienna in June 1905, was the first Botanical Congress that actually functioned internationally in character and discussed extensively the botanical nomenclature. The plans for serious study of the nomenclature were carried out in Paris (1900) and a five-year period was given to study the merits and weaknesses of nomenclature. Unlike the Paris Code of 1867 where no special effort was made to insure a voting membership of representative of botanical strength on an international scale, nor to insure that, the membership be limited to competent and designated delegates, the voting membership of Vienna Congress was restricted to its representatives. In Vienna Congress the rules of Paris Code (1867) was slightly modified and amplified considerably.

Some of the important principles included in this Code are as follows:

- The *Species plantarum* (1753) of Linnaeus was regarded as the starting point for naming vascular plants.
- The acceptance of the descriptions of genera as given by Linnaeus in his 5th edition of the *Genera plantarum* 1754 to validate the publication of the generic names appeared in *Species plantarum* (1753).

- The establishment of *Nomina Generica Conservanda*, means that, the generic name having a wide use would be conserved as valid. A list of generic names to be conserved was included.
- Banning of tautonyms.
- Name of new taxa should be accompanied with Latin diagnosis.
- Vienna Code rejected both the Kew Rule and Rochester Rule, but retained the principle of *nomina conservanda* of Rochester Rule and provided a list of it.
- The determination of priority of a name only by the date of publication

The prescription of Vienna Code was divided into principles, rules and the recommendations. The principles are the basic foundation stones of the rules and recommendations, whereas the rules govern the names of nomenclature and anything contrary to the rules cannot be retained. The recommendations are considered as the secondary point for greater uniformity and clarity of nomenclature, which may or may not be accepted. The Vienna Rule was not accepted by most of the workers due to the establishment of *Nomina Generica Conservanda*.

4. **American Code (1907)** The members of the Rochester Code have failed to include their principles in the revision made at Vienna. Hence, the Americans who were not satisfied with the rules adopted by the Vienna Congress, refused to accept the new rules proposed at Vienna. As a result, the dissatisfied American adherents of Rochester Code, published the new American Code with a slight modification of Rochester Code in 1907, under the heading of American Code.

But the American Code was based on most of the proposals of the Rochester Code and agreed to some respects with Vienna Code. The two Were not so different except that the American Code was modified in some aspect to match with Vienna Code. The Code was supported by the adherents of the Rochester Code and gained no more confidence. However, the Committee on Nomenclature of the Botanical Society of America was pressurised by the supporters of the American Code to prepare a new "*Type-based Code*" in 1918, making a significant concession for the supporters that 'a list of *Nomina Generica Conservanda*' might be added if desired. But no such list was prepared and the Code was never adopted by the society. This Code differs from the American Code (Type-based Code) only in certain points.

The American Code differs from the Vienna Code in the following parameters:

- The first provision of the American Code was that, it did not accept the *Nomina Generica Conservanda*.
- It refused to accept Latin diagnosis to accompany the name of new taxa.
- The type concept was accepted.
- It rejected the repetition of the same binomial for a plant, if it was employed previously even if it was illegitimate. These rules never gained international acceptance.

5. **Brussels Code (1912)** The Fourth International Botanical Congress was met in Brussels in 1910. The most significant decision made at Brussels Congress was the establishment of different starting point for the priority of names of non-vascular plants, the recognition of value of type concept as proposed by Rochester Code and some phrases of the Vienna Code. The important criteria included in this congress are:

   - Adoption of rules for different starting points for priority of names of non-vascular plants.

- ★ Recognition of the Type concept as envisaged in Rochester Code 1892, and Clarification of several phrases of Vienna Rule of 1905.

6. **Cambridge Code (1935)** The Fifth International Botanical Congress convened in 1930 at Cambridge, England. The efforts made by T.A.Sprague and M.L.Green on one hand and A.S.Hitchcock on other hand were successful in reconciling the basic differences between the Vienna Code and American Code and brought about harmony among the major botanical fractions. The new rules legislated at Cambridge constituted the Code of nomenclature that was truly international in name and in function, which was used by the botanists all over the world.

The provisions approved by both the parties at Cambridge are as follows:

- ★ The Code adopted the type concept as "the application of names of taxonomic groups is determined by means of nomenclatural types".
- ★ There should be a list of *Nomina Generica Conservanda*.
- ★ Tautonyms are not admissible and
- ★ Latin diagnosis is required after January 1st, 1932.

7. **Amsterdam Code (1947)** The Sixth International Botanical Congress was held at Amsterdam in 1935. A few major changes in the rules were made and it was resolved that "From January 1^{st,} 1935, names of new groups of recent plants, except the Bacteria are considered as validly published only when they are accompanied by a Latin diagnosis". An attempt to select a list of *Nomina specifica conservanda* was thwarted by an overwhelming vote.

The rule of nomenclature as divided into chapters and sections, which in turn were subdivided into Articles and

Recommendations. The articles were accepted as mandatory and binding and the recommendations were provided for guidance. A list of *Nomina Generica Conservanda* was also included in the Code. A supplement to this Code was compiled by T.A. Sprague in 1950, subjected to discussion and resulted in two major decisions:

- Rejection of the establishment of a list of *Nomina Specifica Conservanda* by 208 to 61 votes.
- Shifting of the dates from Jan.1st, 1932 to Jan. 1st, 1935 as the date after which all diagnoses of plants new to Science should be in Latin.

8. **Stockholm Code (1952)** The Seventh International Botanical Congress was held in Stockholm in 1950, in which some minor changes and a few additions to the rules were legislated. The Code introduced number of definitions on types. The word *taxon* (pl.taxa) was introduced for the first time to designate any taxonomic group or entity. It was further added that taxon of any rank, below the rank of a species, which include the type of the species to be designated by repeating epithet. Addition of a new article on apomicts and rejection of *Nomina Specifica Conservanda* was adopted. Among other changes the most important one was summarised by J. Lanjouw 1952, Rapporteur. General rules of the Stockholm Congress are as follows:

   a. The term 'taxonomic group' was changed to 'taxon'. (pl. taxa).

   b. Publication date of work serving as starting point will be fixed either on Jan 1st or Dec. 31st.

   c. The term 'legitimate, illegitimate, effective and valid' are defined under a new article.

   d. Instead of article 18, dealing with the subject of selecting and designating types, five new articles and two recommendations are accepted.

e. The Congress rejected the principle of *Nomina Specifica Conservanda* by a ballot of 320 votes to 40.

f. The date of effective publication is the moment when the publication becomes available. Name published on labels of exsiccatae are treated as not validly published.

g. Name of species or taxa below the rank of species are regarded as not validly published, if the generic name which forms a part of the binomial has not been validly published before or at the same time.

h. It is recommended to avoid the use of names previously published as 'nomina nuda'.

i. In case of publication of new taxa by one author in the work of another author, the two authors names should be connected by the word 'in'.

j. If the rank of a taxon above that of a genus is changed, the root of the name must be preserved, but ending only has to be changed.

k. Alternative name shall be considered as illegitimate.

l. It is recommended that the publication of new taxon must be accompanied by a figure.

m. A name, of which the type has been excluded, can only be maintained by 'conservation'. In this case the author of the name is the botanist who gives the new description.

n. Misidentification must not be cited in the synonymy. a misapplied name is indicated by the words *'auct non'* followed by the name of the original author and the bibliographical refrence of the misidentification.

o. If the status of the taxon bearing a binary name is changed from that of a species to that of a hybrid or inversely, the name of the original author will be maintained with the addition between brackets of 'pro sp' or 'pro hybr'.

p. In article 62, the sentence dealing with a list of *nomina ambigua* is deleted.

q. It is recommended to write the epithets of species of taxa of lower rank with a small initial.

r. Under the new rule of typification, the application of names of taxa is determined by means of nomenclatural type. A nomenclatural type is that constituent element of a taxon to which the name of the taxon is permanently attached, whether as an accepted name or a synonym.

9. Paris Code (1956) The Eighth International Botanical Congress was held in Paris in July, 1954. The general procedure followed in this Code was very similar to that followed for the Stockholm Code (1952). It greatly emphasized on types (Appendix IV), but the rule of Latin came under fire. In the Congress it was decided that the Code should be published in English, French and in German languages. One of the main changes made and considered as the major feature of the Code in this session was, the separation of the Preamble and Principles from the Rules and Recommendations. Appendix III with *Nomina Generica Conservanda et Rejicienda* was amended and supplemented. Determinations of type (Appendix IV) is reprinted as verbatim in accordance with the definite decision to do so, taken by the Nomenclature Section at Paris.

The important decisions of this Congress include:

★ The adoption of a list of *Nomina Specifica Rejicienda* which has the same stabilizing effect as introducing a *Nomina Specifica Conservanda*.

★ The Code was divided into four main divisions:

i. General considerations

ii. Rules and Recommendations

iii. Provisions for the modification of the Code

iv. Appendices

- To the guiding principles as stated in chapter 1 of the Stockholm Code, two more principles have been added. One stating that 'the application of the names of taxonomic group is determined by means of nomenclatural type' and the other stressing 'the correct use of publications'.
- Rules were proposed for effective publications, name of subdivisions of genera, transference of names between plant and animal kingdoms.
- Proposals to delete article 35 were rejected.

10. **Montreal Code (1961)** The Ninth International Botanical Congress was held at Montreal, Canada in August, 1959.In general, the procedure adopted in the preparation of the Montreal Code was similar to that of Paris and Stockholm Codes. The Congress appointed a special committee to study the question of possible. *Nomina Specifica Conservanda* was not accepted, but it was agreed that an attempt was made to assess the scope and importance of the problem before changing the rules. The main structure of the Code was not changed to any considerable degree. The committee decided to retain as far as possible the numbers of the Articles and Recommendations of the Paris Code. However, some of the examples in the previous Code was not accurate or suitable and it was deleted. Hybrid names and some special categories were slightly amended. Special provisions concerning the fossil plants (Appendix II) found in Paris Code was not incorporated in the Code, but it was agreed to include these ' Special provisions' where necessary in the body of the Code. In this Code,

it was made clear that naming of fossil plants should follow the same lines as that of the recent ones. In the Montreal Code, *Nomina Familiarum Conservanda* for Angiosperms Appendix II was introduced.

11. Edinburgh Code (1966) The Tenth International Botanical Congress was conducted at Edinburgh in August 1964 and in which the report of the special Committee appointed during the ninth Congress was submitted. The numbers of articles were not changed from those of Paris and Montreal Code. A few changes occurred in the numbering of Recommendations mainly due to some deletions. The nomenclature of bi-generic and multi-generic hybrids was amended. *Nomina Familiarum Conservanda* was made and family names of recently discovered Gymnosperms were also added. Further a Committee was set up by the Edinburgh Congress to prepare a glossary of the main technical terms used in the Code, which resulted in '*An Annotated Glossary of Botanical Nomenclature*' (Mc Vaugh *et al.*, 1968).

Some of the important extracts of the Code are as follows:

- A.L.de Jussieu's *Genera Plantarum* (1789) is considered as the starting- point for family names. Although, many of the names included in the list are not a matter of dispute, but their inclusion in the list forms a ready reference providing correct name of the families together with a type genus for each family.
- In the list of *Nomina Familiarum Conservanda*, a few names have spelling somewhat different from long-established names, e.g., Cannabaceae for Cannabinaceae, Capparaceae for Capparidaceae, Haloragaceae for Haloragidaceae, Melastomataceae for Melastomaceae etc.

- ★ The rule of Latin for new diagnoses was finally settled and no amendments were proposed.

12. Seattle Code (1972) The Seattle Code was proposed in the Eleventh International Botanical Congress held in Seattle in August, 1969. This Code was edited by F.A.Stafleu and published in 1972. Most of the proposals submitted to the Congress were concerned with refinements and increased precision rather than with sweeping changes. Some amendments in the list of *Nomina Conservanda* were made. Different Committees were set-up to solve some enquiries regarding the status of names of taxa above the rank of family and the applications of Articles 69, 70 and 71 for incorporation in the Leningrad Code of 1975. The main discussion at Seattle Congress as the tautonymous designation of taxa between genus and species and that below species, a perennial question of superfluous names and a major reorganization of the rules of hybrids. Articles 22 and 26 were refined and the word autonym (name established automatically) was introduced in this Code. Most of the proposals submitted in this Congress were concerned with refinements and increased precision.

The chief issues included in Seattle Code are as follows:

- ★ Due importance was given to priority of publication.
- ★ Adoption of a list of nomina conservanda.
- ★ Each taxonomic group with particular circumscription and rank must bear only one name.
- ★ The tautonymous (Name in which the specific epithet exactly repeats the generic name) designations of taxa between genus and species and that below species.
- ★ All scientific names are treated in Latin, irrespective of their derivation from any source.

- ★ Recognition of superfluous names and separate rule for giving names to the hybrids have been formulated.
- ★ The rank of taxa in ascending order are species, genus, family, order and classes. When greater number of taxa are to be incorporated, additional taxa are constructed by adding the prefix sub to terms denoting the ranks of the individual.
- ★ Legitimate name of the epithet is the one which is suggested according to the rules of nomenclature.
- ★ The word autonym (automatically established names) was introduced in this Code.
- ★ Names suggested in contravention of the rule is termed as illegitimate.

13. Leningrad Code (1978) The Twelth International Botanical Congress was met in Leningrad, USSR (Russia) in July, 1975 and its recommendations came out only in the year 1978. The Code was prepared and edited under the chairmanship of F.A.Stafleu. This current Code shows only small a difference from the Seattle Code. The numbering of Articles and Recommendation is the same as that of the previous Code. Therefore Leningrad Code indicates only few differences from the Seattle Code and the changes made in the Code are as follows:

- ★ The concept of organ- genera was eliminated for fossil plant (Art.3).
- ★ Individual paragraphs on all Articles and Recommendations are numbered in a decimal-like system, and some paragraphs are rearranged.
- ★ The Code makes it clear that it should not apply to names of organisms treated as bacteria and it applies to all other organisms treated as plants (Preamble 7, Principle I, Art.45.4 and Art.65).

- ★ The principle of automatic typification was extended to those names of taxa above the rank of family that are ultimately based on the generic names, however, on priority basis (Art.16, Rec.16A and 16B).
- ★ A new name or combination published before 1953, without indicating the rank, is considered validly published but imperative in question of priority except for homonym and for certain names to be accepted at the varietal rank (Art.35).
- ★ The Article 69 of the previous Code was modified as 'a name to be rejected if it has been widely and persistently used for a taxon not including its type.
- ★ The Articles 70 and 71 of the Seattle Code dealing with discordant elements and monstrosities were deleted but the Article numbers are retained as such to facilitate the use of the Code.
- ★ A thorough rewriting of the section on Orthography is presented (Art.73, Rec.73C, Rec 73G and Art.75).

14. **Sydney Code (1983)** The Sydney Code was based upon the decisions made by Nomenclature Section of the Thirteenth International Botanical Congress held in August 1981at Sydney, Australia. It aimed at the provision of a stable method of naming taxonomic groups, avoiding and rejecting the use of names which may cause error or ambiguity or throw science into confusion. Next importance is given to avoid the useless creation of names. The following inclusion such as absolute grammatical correctness, regularity or euphony of names, more or less prevailing custom, regard for persons etc. are the other consideration given in the Code not withstanding their undeniable importance are relatively accessory.

The Sydney Code is quite similar to the Leningrad Code except for a few minor changes which are enumerated below:

a. More than one hundred proposals were accepted by Sydney Congress merely refining Leningrad Code.

b. Conservation of names of species was accepted but it was not included in the Code.

c. Numbering scheme introduced in this Code makes the reference more simple.

d. Article 59 and related provisions were thoroughly remodelled and a change in starting point date in fungi was adopted.

e. The Articles 60 and 61 were remodelled.

f. The Article dealing with spelling as used by Linnaeus (Article 74) was deleted at Sydney.

g. Some of the dates for the application of various rules were eliminated at Sydney Congress.

h. The provisions for hybrids were thoroughly overhauled.

i. Autonyms were treated as validly published having priority over the names of the same date automatically established.

j. *Nomina Generica Conservanda et Rejicienda* (Appendix III) was considerably revised.

15. Berlin Code (1988) The Fourteenth International Botanical Congress convened at Berlin, Germany in 1987, supersedes all other earlier Codes of Botanical Nomenclature.

The main difference between Berlin Code and the previous editions of the nomenclature is that, the text is published only in English, instead of different languages other than English such as French and German which were already operative since Paris Code of 1956. Otherwise, the Berlin Code is quite similar to Sydney Code. However, one new article and some new recommendations are added in Berlin Code in certain paragraphs some new numbers have been provided.

The *Nomina Specifica Conservanda,* which was accepted but not included in Sydney Code for the first time introduced at this Congress (Appendix IIIB as *Nomina Specifica Conservanda et Rejicienda*). Two species such as *Lycopersicon esculentum* P.Miller and *Triticum aestivum* Linn., have been conserved against the rule of priority, as these names are widely used and any change in these names may cause confusion. In this Congress the two Articles 66 and 67 dealing with illegitimate names were removed. The Recommendation 75A dealing with gender of generic names in Sydney Code appear as a new article (Article76). Articles 7-10 dealing with typification have been revised thoroughly and the '*Guide for the determination of types*' is placed under the recommendation 7B. In addition, rules on types and typification have been incorporated, and also the importance was given to the place of preservation of the type of the name of taxon mentioned from Jan 1st, 1990. The other changes relating to illegitimate, orthography and naming of hybrids are also given in this Code.

The following is the outline of Berlin Code 1988.

Preamble

Division I. Principles (1-6)

Division II. Rules and Recommendations (Articles1-76)

Chapter I. Ranks of taxa and the terms denoting them (Articles 1-5).

Chapter II. Names of taxa (Articles 6-15).

Chapter III. Nomenclature of taxa according to their rank (Articles 16-28).

ChapterIV. Effective and Valid publications (Articles 29-50).

Chapter V. Retention, choice and rejection of names and epithets (Articles 51-72).

Chapter VI. Orthography of names and epithets and gender of generic names (Articles 73-76).

Division III. Provision for modification of the Code are given in four

Appendices and are as follows:

Appendix I. Names of hybrids.

AppendixIIA. Nomina familiarum algarum,fungorum et pteridophytorum.

Appendix IIB. Nomina familiarum bryophytorum et spermatophytorum conservanda.

Appendix IIIA. Nomina generica conservanda et rejicienda.

Appendix IIIB. Nomina specifica conservanda et rejicienda.

*Appendix IV.*Nomina utique rejicienda.

16. Tokyo Code (1994) The Fifteenth International Botanical Congress was held at Yokohama, Tokyo in August-September 1993, under the chairmanship of W.Greuter. This Code was preceded by the Louis Code 2000. The proceedings of Tokyo Code have been translated into Chinese, French, German, Japanese, Russian and Slovak. In this Code extensive renumbering had taken place and therefore, its preface included tabulation, comparing the placement of its provisions with those of Berlin Code.

The following is the outline of Tokyo Code 1994:

Preface Key to the Numbering of the Articles, Notes and Recommendations

Important Dates in the Code

Preamble

Division I Principles

Division II Rules and Recommendations

Chapter I Ranks of taxa (Art. 1, 2, 3, 4, 5)

Chapter II Names of taxa (general provisions)

Section 1 Definitions (Art. 6)

Section 2 Typification (Art. 7, 8, 9, 10)

Section 3 Priority (Art. 11 and 12)

Section 4 Limitation of the principle of priority (Art. 13, 14, 15)

Chapter III Nomenclature of taxa according to their rank

Section 1 Names of taxa above the rank of family (Art. 16, 17)

Section 2 Names of families and subfamilies, tribes and subtribes (Art. 18, 19)

Section 3 Names of genera and subdivisions of genera (Art. 20, 21, 22)

Section 4 Names of species (Art. 23)

Section 5 Names of taxa below the rank of species (infraspecific taxa) (Art. 24, 25, 26, 27)

Section 6 Names of plants in cultivation (Art. 28)

Chapter IV Effective and valid publication

Section 1 Conditions and dates of effective publication (Art. 29, 30, 31)

Section 2 Conditions and dates of valid publication of names (Art. 32, 33, 34, 35, 36, 37, 38, 39, 40, 41, 42, 43, 44, 45)

Section 3 Citation of authors' names for purposes of precision (Art. 46, 47, 48, 49, 50)

Section 4 General recommendations on citation (Rec. 50A-F)

Chapter V Rejection of names (Art. 51, 52, 53, 54, 55, 56, 57, 58)

Chapter VI Names of fungi with a pleomorphic life cycle (Art. 59)

Chapter VII Orthography of names and epithets and gender of generic names

Section 1 Orthography of names and epithets (Art. 60 and 61)

Section 2 Gender of generic names (Art. 62)

Division III Provisions for modification of the Code

Appendix I Names of hybrids (Art. H.1, H.2, H.3, H.4, H.5, H.6, H.7, H.8, H.9, H.10, H.11, H.12)

Appendix IIA Nomina familiarum algarum, fungorum et pteridophytorum conservanda et rejicienda

Appendix IIB Nomina familiarum bryophytorum et spermatophytorum conservanda

Appendix IIIA Nomina generica conservanda et rejicienda

Appendix IIIB Nomina specifica conservanda et rejicienda

Appendix IV Nomina utique rejicienda (B. Fungi, C. Bryophyta, E. Spermatophyta)

Appendix V Opera utique oppressa

17. **Saint Louis Code (2000)** The proceedings of the Sixteenth International Botanical Congress was held in July-August, 2000 at Saint Louis, Missouri, U.S.A. under the chairmanship of Prof. Werner Greuter. It is written entirely in English (British). The Code published by Greuter *et al* in 2000 is called Saint Louis Code. The St. Louis Code was translated in Chinese, French, Japanese, Portuguese, Russian, Slovak and in Spanish.

Louis Code has done some minor changes, clarified certain ambiguous terms like Epitype, and deleted some highly controversial provisions like REGISTRATION of new names etc. The Code also added some new NOTES and EXAMPLES and in some cases the articles are re-numbered and brought together under rules involving typification and deleted some duplicated examples.

The basic structure of the Louis Code remains unaltered like that of Tokyo Code but with some minor changes. The detailed provision of the Code is divided into Rules, set out in Articles, and Recommendations. The object of the "Rule" is to put the nomenclature of the past in order and to provide for future. Plant names on the contrary to a rule cannot be maintained. The Recommendations deal with subsidiary points. Their object is to bring about great uniformity and clearness, especially in future nomenclature. Names are contrary to Recommendation cannot be rejected, but they are not the correct examples to be followed. The Rules and Recommendations apply to all organisms whether fossil or non- fossil, including fungi but excluding bacteria. For bacteria, there is a separate Code which was proposed to be called as International Code of Nomenclature of Bacteria (ICNB) to govern the bacterial nomenclature.

A broad outline of the Saint Louis Code is listed below:

Preamble

Division I. This division includes 6 principles, which form the basis of the system of Botanical Nomenclature.

These are:

I. Botanical nomenclature is independent of Zoological and Bacteriological nomenclature. The Code applies equally to names of taxonomic groups treated as plants whether or not these groups were originally so treated.

II. The application of names of the taxonomic groups is determined by means of nomenclatural types.

III. The nomenclature of a taxonomic group is based up on priority of publication.

IV. Each taxonomic group with a particular circumscription, position and rank can bear only one correct name, the earliest that is in accordance with the Rules, except in specified cases.

V. Scientific names of taxonomic groups are treated as Latin, regardless of their derivation.

VI. The Rules of Nomenclature are retroactive unless expressly limited.

Division II. Rules and Recommendations (Art.1-62 This division contains the Rules distributed in 62 Articles and Recommendations. The names which contravene (contradict) any one or more of the above Rules are illegitimate. Recommendations deal with subsidiary points. The main aim is to bring about greater uniformity in Nomenclature. Names on the contrary to Recommendations need not be rejected, but are not used as examples in the Code.

Chapter I. Taxa and their ranks (Art. 1-5).

Chapter II. Status, typification, and priority of names (Art.6-15).

Chapter III. Nomenclature of taxa according to their rank (Art.16-28).

Chapter IV. Effective and Valid publication (Art. 29-50).

Chapter V. Rejection of names (Art.51-58).

Chapter VI. Names of fungi with a pleomorphic life cycle (Art.59).

Chapter VII. Orthography and gender of names (Art.60-62).

Division III. Provisions for the Governance of the Code: The provisions for modifications are mentioned in 7 Appendices.

These are:

Appendix I.	Names of hybrids.
*Appendix*IIA.	Nomina Familiarum Algarum, Fungorum, Pteridophytorum et Fossilium Conservanda et Rejicienda.
Appendix IIB.	Nomina Familiarum Bryophytorum et Spermatophytorum Conservanda.
Appendix IIIA.	Nomina Generica Conservanda et Rejicienda.
Appendix IIIB.	Nomina Specifica Conservanda et Rejicienda.
Appendix IV.	Nomina Utique Rejicienda.z
Appendix V.	Opera Utique Oppressa.

18. Vienna Code (2005) The Seventeenth International Botanical Congress was held in Vienna in 2005. This Code supersedes the Saint Louis Code and this Code was prepared and edited under the chairmanship of J. Mcneill. The Vienna Code does not differ substantially in over all presentation and arrangement from the St. Louis Code and the numbering of Articles remains the same with a few additions and modifications of paragraphs. One small change has been made in the numbering of the Appendices to make it more logical: the former Appendix IIIA, dealing with conserved names of genera is now considered simply Appendix III, and the former Appendix IIIB with names of species becomes Appendix IV. With Appendix IIA and IIB continuing to contain the two sorts of conserved family names, there is now a logical sequence for the lists of conserved names:

II for families, III for genera, and IV for species. The subsequent Appendices increase in number accordingly, so that names rejected 'utique' under Art.56 form App. V, and suppressed works, App. VI. The "Important Dates in the Code" which was omitted by St. Louis Code has been restored in the Vienna Code from a draft by D.L. Hawksworth.

The Code included Glossary for the first time which appears as Appendix VII. The draft was prepared by P.C.Silva to whom the Committee was grateful. The text of the Code uses three different sizes of print for the Recommendations and Notes being set in smaller type than the Articles, Examples and footnotes are less than the Recommendations and Notes. This type size reflect the distinction between the rules which are mandatory (Articles), Complementary information or advice (Notes and Recommendations) and explanatory material (Examples and footnotes). As in previous edition, scientific names under the jurisdiction of the Code, irrespective of rank are printed in *italic type.* Several proposals on criteria for valid publication of names were considered in Vienna. It was made explicit that names be composed only of letters of the Latin alphabet except as otherwise provided in the Code. Article 33, dealing with new combination was again clarified, principally in making a separation in the paragraphs of the Article between the situation before 1st Jan. 1953 and the more precise requirements from that date onward. In addition it was made clear that, prior to 1st January 1953, when the epithet of a previously and validly published name that applies to the same taxon is adopted, the "presumed new combination" is validly published if there is any indication at all of a basionym however if there is no such indication, the new combination is only validly published, if it would otherwise be a validly published name. By contrast, it was accepted in Vienna that on or after 1st January 1953 a claimed new combination or avowed substitute, that lacks the full information required regarding the basionym or replaced synonym is not validly published even though the name would otherwise be validly published as the name of a new taxon. Three important sets of changes were accepted in Vienna applying to names in particular groups of organisms, fossil plants, pleomorphic fungi, and fungi that had previously been named under the International Code of Zoological Nomenclature (ICZN), respectively. In this code it was made clear that, later homonyms are illegitimate whether the type is fossil or non-fossil. The Congress agreed to move the date (illustrations permitted as types of names published before 1st January 1958, but were prohibited thereafter unless it were " impossible to preserve a Specimen") and decided that for names of microscopic algae and microfungi for which preservation of a type was technically difficult, the type might be an illustration, but for all other organisms, names published on or after 1st January 2007 would require a specimen as type. One further date limit first appears in the Vienna Code. From 1st January 2007 a new combination, a new generic name with a basionym, or an avowed substitute is not validly published unless its basionym or replaced synonym is cited. Currently, although a full and direct reference to the place of publication must be given, the basionym or replaced synonym need to be indicated. The Vienna Code includes the following :

TITLE

Preface Important dates

Preamble

Division I Principles (I-VI)

Division II Rules and Recommendations (Art. 1-62)

Chapter I Taxa and their ranks (Art. 1, 2, 3, 4, 5)

Chapter II Status, typification, and priority of names(Art 6-15)

Section 1 Status definitions (Art 6)

Section 2 Typification (Art. 7, 8, 9, 10

Section 3 Priority (Art. 11 and 12)

Section 4 Limitation of the principle of priority (Art. 13, 14, 15)

Chapter III Nomenclature of taxa according to their rank (Art. 16-28)

Section 1 Names of taxa above the rank of family (Art. 16 and 17)

Section 2 Names of families and subfamilies, tribes and subtribes (Art. 18 and 19)

Section 3 Names of genera and subdivisions of genera (Art. 20, 21, 22)

Section 4 Names of species (Art. 23)

Section 5 Names of taxa below the rank of species (infraspecific taxa) (Art. 24, 25, 26, 27)

Section 6 Names of plants in cultivation (Art. 28)

Chapter IV Effective and valid publication (Art. 29-50)

Section 1 Conditions and dates of effective publication (Art. 29, 30, 31)

Section 2 Conditions and dates of valid publication of names (Art. 32, 33, 34, 35, 36, 37, 38, 39, 40, 41, 42, 43, 44, 45)

Section 3 Author citations (Art. 46, 47, 48, 49, 50

Section 4 General recommendations on citation (Rec. 50A-F)

Chapter V Rejection of names (Art. 51, 52, 53, 54, 55, 56, 57, 58)

Chapter VI Names of fungi with a pleomorphic life cycle (Art. 59)

Chapter VII Orthography and gender of names (Art. 60-62)

Section 1 Orthography (Art. 60 and 61)

Section 2 Gender (Art. 62)

Division III Provisions for the governance of the Code

Appendix I Names of hybrids (Art. H.1, H.2, H.3, H.4, H.5, H.6, H.7, H.8, H.9, H.10, H.11, H.12)

Appendix II A Nomina familiarum algarum, fungorum, pteridophytorum et fossilium conservanda et rejicienda

A. Algae
B. Fungi
D. Pteridophyta

F. Fossil plants (excl. diatoms)

*Appendix*IIB Nomina familiarum bryophytorum et spermatophytorum conservanda

C. Bryophyta
E. Spermatophyta

Appendix III Nomina generica conservanda et rejicienda

A. Algae

A1. Bacillariophyceae (incl. fossil diatoms)

A2. Bodonophyceae

A3. Chlorophyceae

A4. Chrysophyceae

A5. Cyanophyceae

A6. Dinophyceae

A7. Euglenophyceae

A8. Phaeophyceae

A9. Rhodophyceae

A10. Trichomonadophyceae

A11. Xanthophyceae

B. Fungi

C. Bryophyta

C1. Hepaticae

C2. Musci

D. Pteridophyta

E. Spermatophyta

E1. Gymnospermae

E2. Monocotyledones

E3. Dicotyledones

F. Fossil plants (excl. diatoms)

Appendix IV Nomina specifica conservanda et rejicienda

A. Algae

B. Fungi

C. Bryophyta

D. Pteridophyta

E. Spermatophyta

F. Fossil plants (excl. Diatoms)

Appendix V Nomina utique rejicienda

A. Algae

B. Fungi

C. Bryophyta

D. Pteridophyta

E. Spermatophyta

Appendix VI Opera utique oppressa

Appendix VII Glossary of terms used and defined in this Code

19. Melbourne Code (2011) The Eighteenth International Botanical Congress was held in Melbourne, Australia in July 2011 and the rules are yet to be published.

MELBOURNE CODE (2011) The Eighteenth International Botanical Congress was held under the chairmanship of J. MCNEILL in Melbourne, Australia in July, 2011. The ***International Code of Nomenclature for algae, fungi, and plants*** (***ICN***) is the set of rules and recommendations dealing with the formal botanical names that are given to plants, fungi and a few other groups of organisms, all those "traditionally treated as algae, fungi, or plants". It was formerly called the ***International Code of Botanical Nomenclature*** (ICBN); this name was changed at the International Botanical Congress in Melbourne in July 2011 as part of the Melbourne Code which replaces the Vienna Code of 2005.

For naming of cultivated plants there is a separate code, the **International Code of Nomenclature for Cultivated Plants (ICNCP),** which gives rules and recommendations that supplement the **ICN.** The changes to the **International Code of Nomenclature for algae, fungi, and plants** (**ICN**) are decided on every 6 years at Nomenclature Sections associated with International Botanical Congresses (IBC). The **ICN** can only be changed by an International Botanical Congress (IBC), with the **International Association for Plant Taxonomy (IAPT)** providing the supporting infrastructure. Each new edition supersedes the earlier editions and is retroactive back to 1753, except where different starting dates are specified. Melbourne Code (2011) supersedes the earlier Vienna Code of 2005, published six years ago in Vienna, Austria.A number of significant changes were made to the Melbourne Code in the rules governing the botanical nomenclature.

Some of the important features of the code are as follows:

- There is a change in the title of the name of Code, i.e. The name **International Code of Botanical Nomenclature (ICBN)** was changed to **International Code of Nomenclature for algae, fungi, and plants,** in Melbourne . (In Melbourne mycologists felt that, the word "**Botanical**" was misleading and could imply that the Code covered only green plants and excluded fungi and diverse algal lineages, hence, it was agreed that the name be changed to ***International Code of Nomenclature for algae, fungi, and plants.*** In referring to the Code under its new title, the abbreviation **ICN** will be used and it takes effect as soon as ratified by the congress (on Saturday 23rd July, 2011).
- For the first time in history the Code now permits only electronic - publication of names of new taxa (online in Portable Document Format (PDF) with an International Standard Serial Number (ISSN) or an International Standard Book Number (ISBN)) and it effects from 1st January, 2012; no longer will it be a requirement to deposit some paper copies in libraries.
- The content of a particular electronic publication must not be altered after it is first issued. Any such alterations are not themselves effectively published. Corrections or revisions must be issued separately to be effectively published.
- An electronic publication is not effectively published if there is evidence associated with or within the publication that it is merely a preliminary version that was, or is to be, replaced by a version that the publisher considers final, in which case only that final version is effectively published.

- The Latin description or diagnosis required for valid publication of the name of a new taxon of all organisms falling under the Code may be either in English or Latin with the effect from 1st January, 2012.
- In addition, new names of organisms treated as fungi, in order to be validly published, from 1st January, 2013, must include in the protologue (everything associated with a name at its valid publication) the citation of an identifier issued by "a recognized repository" (such as MycoBank and Index Fungorum).
- "One fungus, one name" and "one fossil, one name" are important changes for fungi and for fossils; the concepts of *anamorph* and *teleomorph* (for fungi) as well as *morphotaxa* (for fossils) have been eliminated.
- The requirement for a unique identifier for new names of fungi from 1st January, 2013 does not apply to plants or algae; there is no need for authors of new names in these groups to request Life Science Identifiers (LSIDs) - or other identifiers - from indexing centers.
- The Melbourne Code is written entirely in (British) English; Like the Vienna Code, it is also (Melbourne Code translated into several languages.
- The title of the Code was broadened as mentioned above to make explicit that it applies not only to plants, but also to algae and fungi.
- The volume of Melbourne Code comprises only the main text of the Code, that is the Preamble, Division I Principles, Division II Rules and Recommendations, Division III Provisions for the Governance of the *Code*, Appendix I Names of Hybrids, the Glossary,

the Index of scientific names, and the Subject index (other Appendices published only electronically).

- Authors should preferably publish in publications that are archived, satisfying the following criteria as far as is practical (see also Rec. 29A.1):
  a. The material should be placed in multiple trusted online digital repositories, e.g. an ISO-certified repository;
  b. Digital repositories should be in more than one area of the world and preferably on different continents;
  c. Deposition of printed copies in libraries in more than one area of the world and preferably on different continents is also advisable.
- To aid availability through time and place, authors publishing nomenclatural novelties should give preference to periodicals that regularly publish taxonomic articles. Otherwise, a copy of a publication (whether published as printed or electronic matter) should be sent to an indexing centre appropriate to the taxonomic group, and publications that exist only as printed matter should be deposited in at least ten, but preferably more, botanical or other generally accessible libraries throughout the world.
- The date of effective publication is the date on which the printed or electronic matter became available to the publishers and in the absence of proof establishing some other date, the one appearing in the printed or electronic matter must be accepted as correct.(31.1)
- When a publication is issued in parallel electronic and printed versions, these must be treated as effectively published on the same date unless the

dates of the versions are different according to Article 31.1.

★ The Glossary, a new feature in the Vienna Code, has retained its basic structure but has been revised and updated with the new entries which include: "author citation", "binding decision", "element", "isoepitype", "isolectotype", "isoneotype", "name of a new taxon", "organism", "suppressed works", and "type designation", while some existing entries have been substantially revised, e.g. "basionym", and "confusingly similar names"; for others, such as "name at new rank (status novus)" and "replacement name (avowed substitute)", the primary entry has been changed to reflect the preferred term in the Code. Five entries have been deleted ("exsiccata", "form taxon", "holomorph", "morphotaxon", and "plant"), reflecting the fact that these terms are no longer used in the Code (or not in any special way). This reflects the role of the Glossary which is strictly to explain terms used in the Code, and where possible to do so using the precise wording associated with these terms in the Code.

MAIN OUTLINE OF MELBOURNE CODE 2011

Preface Key to the re-numbering of Articles, Notes, and Recommendations

1. Vienna Code to Melbourne Code
2. Melbourne Code to Vienna Code

Important dates in the Code

Preamble

DIVISION I. Principles

DIVISION II. Rules and Recommendations

CHAPTER I. Taxa and their ranks Art. 1, Art. 2, Art. 3, Art. 4, Art. 5

CHAPTER II. Status, typification, and priority of names

Section 1 Status definitions Art. 6

Section 2 Typification Art. 7, Art. 8, Art. 9, Art. 10

Section 3. Priority Art. 11, Art. 12

Section 4. Limitation of the principle of priority Art. 13, Art. 14, Art. 15

CHAPTER III Nomenclature of taxa according to their rank

Section 1. Names of taxa above the rank of family Art. 16, Art. 17

Section 2. Names of families and subfamilies, tribes and subtribes Art. 18, Art. 19

Section 3. Names of genera and subdivisions of genera Art. 20, Art. 21, Art. 22

Section 4. Names of species Art. 23

Section 5. Names of taxa below the rank of species (infraspecific taxa) Art. 24, Art. 25, Art. 26, Art. 27

Section 6. Names of organisms in cultivation Art. 28

CHAPTER IV Effective publication

Section 1 Conditions of effective publication Art. 29, Art. 30

Section 2 Dates of effective publication Art. 31

CHAPTER V Valid publication of names

Section 1 General provisions Art. 32, Art. 33, Art. 34, Art. 35, Art. 36, Art. 37

Section 2 Names of new taxa Art. 38, Art. 39, Art. 40

Section 3 New combinations, names at new ranks, replacement names Art. 41

SHENZHEN CODE 2017 The 19th International Botanical Congress will be held in Shenzhen, China in the year 2017, after the meeting of the event organizing committee. This is not the only time that the IBC will be held in China, and its also the first time the congress ever held in a developing country. Around 6,000 participants are expected to be at the event to hold discussions, exchange views on the issues concerning plant and plant sciences.

6

MAIN OUTLINE OF SAINT LOUIS CODE (2000)

Keywords Major divisions, principles of the code, rules, recommendations, chapters, sections, articles.

The St. Louis Botanical Code (2000) is divided into three major divisions with an introductory note (Preamble). The divisions are as follows:

Division I It includes six principles (I-VI), which form the basis of all the rules legislated at various Congress and are solely aimed at stabilizing the nomenclatural practices. The six principles of the Code are as follows:

I. Botanical nomenclature is independent of zoological nomenclature. The Code applies equally to names of taxonomic groups treated as plants whether or not these groups were originally so treated.

II. The application of names of taxonomic groups is determined by means of nomenclatural types.

III. The nomenclature of a taxonomic group is based upon priority of publication.

IV. Each taxonomic group with a particular circumscription, position and rank can bear only one correct name, the earliest that is in accordance with the Rules, except in specified cases.

V. Scientific names of taxonomic groups are treated as Latin, regardless of their derivation.

VI. The Rules of nomenclature are retroactive unless expressly limited.

The principles do not give detailed rules about nomenclature but provide the main ideas of the rules of the ICBN.

Division II It includes Rules and Recommendations. In the beginning, the rules are distributed under 75 Articles and at the Berlin Code (1988), the article number is increased to 76, which are distributed under seven chapters. But in the recent Code some of these article numbers are modified on the basis of type method Art.69 and Art.70-71 dealing with discordant elements and monstrosities were deleted,and Art.66 and 67 with illegitimate names were also removed, but the article numbers are retained to facilitate the use of the Code.

The Rules or Articles give detailed description to put the nomenclature of the past in order and provide for future maintenance of nomenclature. The Rules of nomenclature are divided into chapters and sections. Sections are further divided into Articles, and recommendations. The rules are mandatory and binding. The names that are contrary to the rules cannot be maintained.

In the beginning the Code had only six chapters, after that it was increased to seven, with different sections for various articles and is established for providing a stable method for naming of taxonomic groups avoiding and rejecting the incorrect names applied to taxa which may create error or confusion.

The Recommendations are often practical applications of the rules of nomenclature and do not affect the validity of a name. The Recommendations usually deal with subsidiary points. Their main purpose is to bring out greater uniformity and clarity, especially in the future plant nomenclature.

The Rules and Recommendations are applicable to all organisms treated as plants including fungi, and fossil, excluding bacteria.

Division III It includes the provisions for the Governance of the Code. At the end of rules the provisions for different names of hybrids, names of families, genera and species to be conserved under Nomina Conservanda and guide for the determination of types are all kept under several appendices. Earlier Codes contain only three appendices. Later on in the subsequent International Botanical Congress held at various countries the number was increased to four and again at the Fifteenth International Botanical Congress held at Tokyo (1994), a small change has been made in the numbering of the Appendix to make it more logical, as a result, the Appendix number has increased to five and the same number has been followed in the subsequent Code. Each Appendix includes a special topic and is a source of instructional guidance and also provides supplementary information.

The main outline of the Saint Louis Code (Greuter *et al.*, 1999) is given below:

Preamble (Pre. 1-Pre.11)

Division I Principles (I-VI)

Division II Rules and Recommendations (Art. 1-62)

Chapter I Taxa and their ranks (Art. 1,2,3,4,5)

Chapter II Status, typification, and priority of names (Art.6-15)

Section 1 Status definitions (Art.6)

Section 2 Typification (Art. 7,8,9,10)

Section 3 Priority (Art. 11 and 12)

Section 4 Limitations of the principle of priority (Art. 13,14,15)

Chapter III Nomenclature of taxa according to their rank (Art.16 - 28)

Section 1 Names of taxa above the rank of family (Art.16 and 17)

Section 2 Names of families and subfamilies, tribes and subtribes (Art. 18 and 19)

Section 3 Names of genera and subdivisions of genera(Art. 20, 21, 22)

Section 4 Names of species (Art. 23)

Section 5 Names of taxa below the rank of species(infraspecific taxa) (Art. 24, 25, 26, 27)

Section 6 Names of plants in cultivation (Art. 28)

Chapter IV Effective and Valid publication (Art. 29-50)

Section 1 Conditions and dates of effective publication (Art.29, 30, 31)

Section 2 Conditions and dates of valid publication of names (Art. 32, 33, 34, 35, 36, 37, 38, 39, 40, 41, 42, 43, 44, 45)

Section 3 Author citations (Art. 46, 47, 48, 49, 50)

Section 4 General recommendations on citation (Rec. 50A-F)

Chapter V Rejection of names (Art. 51, 52, 53, 54, 55, 56, 57, 58)

Chapter VI Names of fungi with a pleomorphic life cycle (Art. 59)

Chapter VII Orthography and gender of names (Art.60-62)

Section 1 Orthography (Art. 60 and 61)

Section 2 Gender of names (Art.62)

Division III Provisions for the Governance of the Code.

The Provisions are included in seven appendices and are as follows:

Appendix I Names of Hybrids (Art.H.1-H.12)

Appendix IIA NOMINA FAMILIARUM ALGARUM, FUNGORUM, PTERIDOPHYTORUM ET FOSSILIUM CONSERVANDA ET REJICIENDA

Appendix IIB NOMINA FAMILIARUM BRYOPHYTORUM ET SPERMATOPHYTORUM CONSERVANDA

Appendix IIIA NOMINA GENERICA CONSERVANDA ET REJICIENDA

Appendix IIIB NOMINA SPECIFICA CONSERVANDA ET REJICIENDA

Appendix IV NOMINA UTIQUE REJICIENDA

Appendix V OPERA UTIQUE OPPRESSA

7

RANKS AND NOMENCLATURE OF TAXA

Keywords Taxon, taxonomic hierarchy, categories, classification, hierarchical ranks, nesting box system, dendrogram, general rule, family names, generic names, specific epithets, hybrids, cultivars, form-genera, examples.

The term 'taxon' has been introduced for the first time in Eighth International Botanical Congress held in Paris in the year 1956. The taxa and their ranks are covered under Articles 1-5 of the Chapter- 1 of the Vienna Code 2005. A "taxonomic group of any rank" is called taxon (Pl. taxa), and the term taxon was proposed by Lam.

Structure of taxonomic hierarchy The system of nomenclature provides a hierarchial arrangement of ranks of taxa in plant classification. Every individual plant is treated as belonging to a number of categories of successively higher rank with the species as the basic unit. The hierarchy applied in the scientific classification of plants is made up of a set of conventional taxonomic terms called categories. The level at which a category stands in the hierarchy is known as its "rank". The hierarchical "categories" and "ranks" are general terms embracing all taxa at certain levels. Taxon of lower rank are always subordinate to and included in, those that are of higher rank.

Botanical classification of taxa involves the assembling of plants into taxonomic groups on the basis of their relationships. Therefore these taxonomic groups are assembled into more inclusive groups, which in turn are assembled into still more inclusive groups. Thus the arrangement of taxa into an ascending series of ever increasing inclusiveness forms, what is known as 'hierarchial system of classification' or 'hierarchy' (ies). Thus the frame work that has been built up for this purpose in the scientific classification of plants is termed as the 'taxonomic hierarchy'.

Classification therefore constitutes categorising a large number of individuals based on their similarities and dissimilarities. In this arrangement the individual plants are kept at the bottom and end up at the top, with an embracing group, the plant kingdom. In between, the various groups of plants at different levels of the taxonomic hierarchy are placed. The taxonomic hierarchies may be illustrated in two different ways such as by a **'plan-view'** or an **'elevation view'**. Both of which show, how one or more taxa, at each level are combined into a single taxon, at the next higher level, and this higher taxon being defined by the sum of the characters of all its subordinate taxa. Theoretically there is no limit to the number of ranks contained in a hierarchy. The International Code of Botanical Nomenclature recognizes twelve principal ranks in the hierarchy and their descending sequence of the categories of different ranks in the hierarchy are Kingdom, Division, or Phylum, Class, Order, Family, Tribe, Genus, Section, Series, Species, Variety and Form (Stace, 1980, Art.3) and except for some fossil plants.

If a greater number of ranks of taxa are required, the terms for these are made either by adding the prefix 'sub' to the terms denoting the ranks or by the introduction of supplementary terms (Art.4). The subcategories generally used are Subdivision, Subclass, Subseries, Suborder, Subgenus, Subspecies etc. Exceptionally Super category like superorder may also be designated by the Code for some of the ranks.

Taxonomic categories of ICBN are presented below

Regnum	Kingdom
Subregnum	Subkingdom
Divisio	Division
Subdivisio	Subdivision
Classis	Class
Subclassis	Subclass
Ordo	Order
Subordo	Suborder
Familia	Family
Subfamilia	Subfamily
Tribus	Tribe
Subtribus	Subtribe
Genus	Genus
Subgenus	Subgenus
Sectio	Section
Subsectio	Subsection
Series	Series
Subseries	Subseries
Species	Species
Subspecies	Subspecies
Varietas	Variety
Subvarietas	Subvariety
Forma	Form
Subforma	Subform
cultivar	(cv) - cultivar

These categories are abstract concepts to which groups of plants are assigned for the purpose of classification. They are employed in a conventional order which must be strictly followed in the same relative position in the hierarchy (Arts. 1-5)

The Code does not recognise the term 'Natural order' which is often used in many of the older books for the modern equivalent of the rank family of the Code. The Code has certain special provisions for naming of the various taxa. Therefore to understand the different categories of the ranks of taxonomic hierarchy, the Code has stipulated certain standardized grammatical endings – (suffix) such as, phyta, phytina, opsida, opsidae, idae, ales, ineae, aceae, oideae, inae etc to indicate its rank for the major categories from division down to subtribe..

(The names of genera are nouns in the singular number, whereas the names of taxa at lower ranks (Species, subspecies, varieties) may be noun or adjectives in the singular or plural, and none is provided with distinctive endings denoting its precise rank)

Hierarchical ranks and their stipulated endings given by ICBN

Rank	Endings	Example
Division	phyta	Magnoliophyta
Subdivision	phytina	Pterophytina
Class	opsida	Magnoliopsida
Subclass	opsidae(lower plants)	Pteropsidae
		Rosidae
	idae (seed plants)	
Order	ales	Rosales

Suborder	ineae	Rosineae
Family	aceae	Rosaceae
Subfamily	oideae	Rosoideae
Tribe	eae	Roseae
Subtribe	inae	Rosinae
Genus	us, -um, -is,	Pyrus, Allium, Arabis,
	-a, -on,	Rosa, Polypogon
Subgenus	Cuscuta	subgenus Eucuscuta
Section	Scrophularia	section Anastomosanthes
Subsection	Scrophularia	subsection Vernales
Series	Scrophularia	series Lateriflorae
Species	Rosa	indica
Subspecies	Crepis sancta subsp	bifida
Variety	Lantana camara var	varia
Form	Tectona grandis f.	punctata

The endings for rank, subclass and above are recommendations, whereas for order and below these are mandatory rules. The names of different groups of ranks above the level of genus are considered as uninomials in the plural case.

For example, all the roses together form one taxon, which is known as *Rosa indica*. This is a species, a group of plants assigned to the taxonomic category species, and standing in the hierarchy at the rank of 'species'. The same way the different species of *Rosa* are grouped together form a larger taxon, known as *Rosa*. This is called genus, which stands at a higher rank in the taxonomic hierarchy, the rank of 'genus'. Above this, the genus *Rosa* and other related genera such as *Rubus*, *Fragaria*, *Pyrus* and *Prunus* together form a still larger taxon of yet higher rank, the family *Rosaceae*. This process which can be graphically expressed as mentioned earlier in the form of plan-view (or bird's – eye view) is often called "**box- within - box**" arrangement or a '**nesting box system**'(Figure 7.1) (Jeffrey, 1968), while the 'elevation view' is otherwise known as '**dendrogram** (Figure 7.2).

Figure 7.1 BOX IN BOX ARRANGEMENT

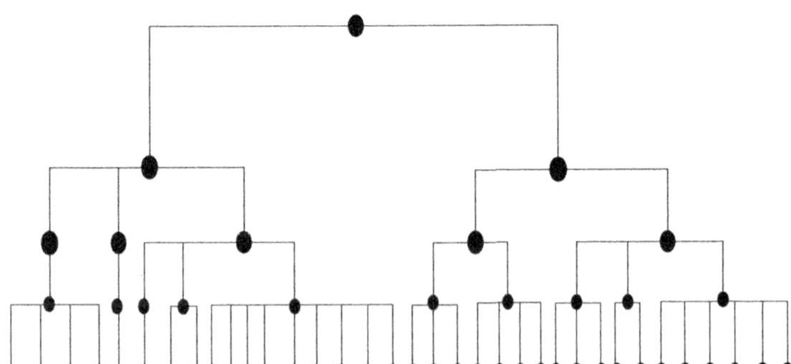

Figure 7. 2 DENDROGRAM

In the above representation, the outermost box represents the 'plant kingdom', the next box inside the kingdom, the 'division', the next box within division denotes 'family'. The next boxwithin family in the diagram indicates the 'genus' (pl.genera) and the smallest boxes within genus represent the 'species'. Species might include still smaller boxes called 'varieties' if any. So also, the families are included in orders, orders under series, series under subclasses, subclasses under classes and classes under division and so on.

In its turn, the family *Rosaceae* together with a few other families, makes up a still higher taxon, the order Rosales; orders in turn are grouped in still higher taxa and the process is repeated until all the embracing taxon of the kingdom is reached.

The use of the hierarchy is also applied to all other plants except the fossil plants. Successive classification into larger and larger groups is the basic feature of the taxonomic hierarchy. Therefore the rank of species forms the basic unit of classification. Hence in the actual practice, one or more of closely related species are grouped into genus, one or more genera into families, families into order, orders into series, series into subclasses, subclasses into classes, classes into divisions and finally divisions make up the plant kingdom.

Family According to Article 18 of the Code, the name of a family must be formed by adding the suffix –aceae to the stem of the name of its type genus. E.g., *Annonaceae* comes from the type genus *Annona*. However, an exception of eight family names of angiosperms, whose family names' ending words are not in accordance with rules of ICBN, and they end with 'ae'. However, the Code sanctioned to use their old names due to their long usage in the literature and at the same time these families also have revised name.

The following is the list of family names with traditional and revised name (Arts.16-27)

Traditional name	Revised name	Type genus of the family
Cruciferae	Brassicaceae	Brassica
Leguminosae	Fabaceae	Faba
Guttiferae	Clusiaceae	Clusia
Umbelliferae	Apiaceae	Apium
Compositae	Asteraceae	Aster
Labiatae	Lamiaceae	Lamium
Palmae	Arecaceae	Areca
Gramineae	Poaceae	Poa

The use of family name *Leguminosae* is sanctioned only as long as it includes all the three subfamilies such as Papilionoideae, Caesalpinioideae and Mimosoideae (Art.52). In case these are considered as separate families, then the name of *Papilionaceae* is conserved against *Leguminosae* for the first of getting the name *Fabaceae* and the name *Fabaceae* is an alternate name, comes from the type genus *Faba* and at present those who regard the family *Papilionaceae*, a name not derived from a generic name as constituting an independent family may use that name, although it is not formed in the prescribed usual manner and the name *Papilionaceae* is also allowed due to the presence of papilionaceous corolla. The name of a family is a plural adjective used as a substantive,. Accordingly any family name should be considered as a plural adjective. Hence the family *Magnoliaceae* are called primitive family and it is wrong to say that the family *Magnoliaceae* is a primitive family. A family is an assemblage of closely allied genera and that resemble each other more closely than with other genera.

Generic name The name of a genus (Art.20) is always uninomial, singular word treated always as a substantive noun. The first letter of the generic name is capitalised. The generic name may be a masculine, feminine or neuter gender.

The gender is usually indicated by the suffix (ending): For example the genera ending with –us, or –pogon commonly refer masculine genera, while the feminine by –a, or –is, and –um for neuter genera.

A genus is an assemblage of a number of species which resemble one another more closely than other species. Genus is a taxon subordinate in rank to family and higher than the species. Each family consists of one or more genera, the family with a single genus is called 'monogeneric family', or 'unigeneric family' as in *Typhaceae* with only one genus *Typha* or *Cannaceae* with the single genus *Canna*.

It is a general rule that, only one generic name is suggested for one plant. Names may be in Latin or may have Latin termination. In case, a generic name consists of two words, they should be joined together by a hyphen. Very long generic names and adjectives must be avoided. A genus may be divided into subgenera and in turn into sections, subsections and series.

The generic name may be taken from any source, what ever and may even be composed, but the name must not coincide with a technical terms used in morphology unless it was published before 1.1.1912.

The common sources for generic names are as follows

a. **Commemoration of a person** Many genera have been named in honour of a patron of Botany or Horticulture or the Botanists. For example, *Hookera* for Hooker, *Grayia* for Asa Gray, *Theophrasta* for Theophrastus, *Dioscorea* for Dioscorides, *Bauhinia* for Bauhin, *Candollea* for Augustin de Candolle, *Magnolia* for Magnol, *Linnaea* for Linnaeus, *Zinobia* for Queen Zenobia of Palmyra, *Jeffersonia* for Thomas Jefferson, *Victoria* for Queen Victoria of England etc. The names honouring a person, whether a man or woman always take the feminine form.

b. **Based on place** Some genera are named based on the place, from where they were collected. For e.g., the genus *Arabis* was collected from Arabia, *Sibiraea* for Siberia, *Araucaria* after Arauco a province of Chile, *Mexicana* for Mexico, *Salvadora* for EL. Salvadore and so on.

c. **Based on salient character** Plant characters are also used in many cases. For example yellow wood in *Zanthoxylon*, *Callicarpa* from the fruits of the Beauty-berry (Beautiful fruits), marshy habit of *Hygrophila, Polygala* with more milky content, liver like leaves in *Hepatica*, *Liriodendron* (lily tree) from the shape of the flowers of the Tulip Tree; *Acanthospermum* with spiny fruit, *Trifolium* with trifoliate leaves, *Oxydendrum* (Sour tree) from the acid leaves of the sourwood etc.

   ★ **Aboriginal names** The generic names are also taken from other than Latin language without alteration of endings. The names existed in the lands where the plants were discovered and afterwards, they were converted into Latin names. For example, the genus *Ginkgo* was discovered from Chinese, *Catalpa* from the Red Indian, *Ailanthus* from the Moluccan, *Tsuga* from the Japanese, *Nelumbo* from the Ceylonese, *Narcissus* a Greek name for daffodils, *Vanda* a Sanskrit aboriginal name, *Ravenala* from the Madagascarian or formed by Latinisation of common names e.g. *Nelumbo*, *Ravenala*, *Tsuga*, *Ailanthus*, *Pandanus* and so on.

d. **Based on mythological or poetic origin** For e.g., *Circaea* (enchanter's night shade) refers to Circe, the enchantress, *Nymphaea* (Water-lilies) denotes the lovely water- nymphs; *Theobroma* (god's food) refers to the Chocolate plant.

The generic name of a tree, whatever be the ending, gets a feminine form since trees are generally considered feminine in classical Latin. *Quercus*, *Pinus* and *Prunus* are the examples of feminine genera.

Specific epithet The name of a species (Art.23) is a binomial, consisting of two words, a generic name followed by a specific epithet.

Species is a taxon subordinate in rank to genus, subgenus, section, subsection or series.

John Ray (1686) defined "the species as a group of plants which breed true from the seeds within its own limits". According to Cronquist (1968) species may be defined as "an inherent taxonomic level, the smallest unit which is permanently distinct and distinguishable".

If the specific epithet consists of two words they should be united or hyphenated. For e.g., *Hibiscus rosa- sinensis*.

The Code recommends that all specific epithets should begin with a lower case initial letter. An upper case initial letter is sometimes used, when the specific epithets are derived from a person's name.. The Code discourages such usage for specific epithets. Specific epithets are often adjectives. Sometimes the specific epithet is a noun when derived from a common Greek or Latin word, e.g., *Capsella Bursa-pastoris* (Bursa-pastoris means Shephards' purse) and such honoured names may be spelled with capital initial under special circumstances. The specific epithets must concur grammatically with the generic name, however it differs in gender.

The masculine gender of specific epithets is indicated by alb**us**, niger, pubescence, repens, japonic**us**, chinensis, vulgaris etc., feminine by alb**a,** nigr**a**, chinense, japonic**a** etc., and neuter by alb**um**, nigrum, japonicum, vulgare etc., A specific epithet may be derived from any source or composed arbitrarily.

The sources commonly used for specific epithets are as follows

i. **Name of person** The specific epithets named after a person may be of genitive (possessive) or adjectival nature.

a. **Genitive form of epithet** The specific epithet when used in genitive form, it takes its form depending on ending the person's name. For e.g., the names ending in a vowel or –er, the letter –i, is added for male person, that is -roylei (*Impatiens roylei*) after Royle, hookeri (*Iris hookeri*) after the person Hooker, but if it ends in a consonant, -ii, is also added for male (wallichii after Wallich), -ae, and –iae for females –toddiae (*Rosa toddiae*) after Todd, wilsoniae after Wilson. If the name ends in –a, then letter –e, is added to the name of the person –senguptae (*Euphorbia senguptae*) after Sengupta.

The names of the persons already in Latin, e.g., Linnaeus, the Latin ending –us in this case has to be dropped before adding the proper genitive ending. E.g., linnaei after Linnaeus. The genitive form of specific epithets are not related to the gender of the genus.

b. **Adjectival form of epithet** The specific epithets when used in adjectival form, it takes its ending from the gender of the genus after adding –ian if the name of a person ends in consonant, adding –an if the name ends in a vowel except when it ends in –a, wherein –n is added.

ii. **Based on place** The specific epithet is also formed by using the name of a place as an adjective. The genus determines ending after the addition of –ian or –ic and then the relevant gender ending 'as' determined by the genus. Examples are given below

Place	Genus	Gender	Specific epithet	Binomial
	Rosa	Feminine	indica	Rosa indica
India	Euonymus	Masculine	indicus	Euonymus indicus
	Solanum	Neuter	indicum	Solanum indicum

The specific epithet may also be formed by adding –ensis (for masculine and feminine gender, e.g., *Hedera napalensis*; *Rubus canadensis*) or –ense (for neuter gender e.g., *Ligustrum nepalense*) to the name of the place. The other examples of specific epithets based on place are *Mangifera indica* after India, *Argemone mexicana* after Mexico, *Iris kashmiriana* for Kashmir, tanjorensis for Tanjore, pulneana for Pulni, etc.

iii. **Based on character** Specific epithets based on an important character of the species are always in an adjectival form and derive their gender from the genus. For example, a name based on the colour of the plant or plant part may be called alba(white) (Eg. *Rosa alba*), album (*Chenopodium album*) or albus (*Melilotus albus*), nigra (black) (*Solanum nigrum*), viridis (green) (*Amaranthus viridis*) and cultivated plants may take the form sativa (*Oryza sativa*), sativum (*Allium sativum*) or sativus (*Lathyrus sativus*) depending on the gender of the genus to which the specific epithet is assigned. The other vital characters such as bicolour (two-coloured), hirta (hairy), grandiflora (large-flowered), cordifolia (cordate-leaved) angustifolia (narrow-leaved), latifolia (broad-leaved), quadrifolia (four leaved) foetida (ill-scented), spinosa (spiny) etc.

iv **Noun in apposition** Sometimes a specific epithet may be a noun carrying its own gender and usually in the nominative case. For e.g., the binomial *Pyrus malus* is based on the Greek name malus, the ancient word for common apple and in *Allium cepa*, 'cepa' is the Latin name for onion, *Capsella Bursa-pastoris* where Bursa-pastoris was the classical name for Shepherd's purse.

v **Constructed noun** The specific epithet in some cases may be constructed from a noun with a suffix, showing affinity or resemblance, e.g. quercifolia (leaves like *Quercus*), amaranthoides (akin to *Amaranthus*), bignonioides (resembling *Bignonia*) etc.

vi **Others** Specific epithets of some names are descriptive adjectives and are generally based on various criteria, which may indicate the following:

a. Habitat of the plant: e.g., aquatica (in water), arvensis (in fields), muralis (on walls), sylvatica (in woods);

b. Utility of the plant: e.g., edulis or esculenta (edible), hortensis (grown in gardens), religiosa (religious);

c. Relative abundance of the plant: e.g., rara (rare), vulgaris (common);

d. Season and duration: e.g., autumnalis or autumnale (of autumn), biennis (biennial),

e. Size, shape or habit of the plant: e.g., alta (tall), gigantea (giant), natans (swimming), pendulus (hanging), repens (creeping), scandens (climbing), procumbens (prostrate);

Both generic name and specific epithet when written or typed should be underlined or when printed the binomial should be in italics or boldface. A generic name, after it has been spelled out in a specific name, if used for other species in continuation, it may be abbreviated using the initial capital, e.g., *Hibiscus vitifolius*, *H. schizopetalous*, *H. esculentus* etc. A specific epithet is usually one word but when consisting of two words they should be hyphenated as in *Trigonella foenum-graecum*.

Intergeneric hybrid The name of intergeneric hybrid is written by connecting the names of the two parent genera with multiplication sign ×.

For e.g. × Agropogon = (Agrostis × Polypogon)

The name of a subdivision of a genus is a combination of a generic name and a subdivisional epithet connected by the terms such as **subgenus, section, series** etc.

E.g., 1. *Mastixia* Blume (1826) subg. *Mastixia*

2. *Saponaria* sect. *Vaccaria*

3. *Aesculus* sect. *Aesculus*

Name of infraspecific categories The name of a taxon below the rank of a species or any category subordinate in ranks to species is called infraspecific taxon (pl.taxa). The International Rules provided for infraspecific categories are subspecies, varietas, subvarietas, forma and so on. The name of an infraspecific taxon is a combination of the name of a species and an infraspecific epithet connected by a term denoting its rank.

The names of subspecies are trinomials and are formed by adding a subspecific epithet to the name of a species of a taxon. The subspecies are regarded as the 'baby species' or species of small magnitude that can be distinguished from each other by minor morphological features.

For e.g. 1. *Crotalaria willdenowia* subsp. *glabrifolia*.

2. *Ammannia baccifera* ssp. *aegyptiaca*.

The variety is a morphological variant of the species. The form (forma) is the smallest category possessing very less variations among the individuals of any population. These variations may be in the corolla colour, fruit colour etc. The variety characters of a species are associated with hereditary characters.

For e.g. 1. *Leucas lavandulifolia* var. *nagalapuramiana.*

2. *Brassica oleracea* var. *capitata.*

A variety (varieta) within a subspecies is called quadrinomial as in *Bupleurum falcatum* ssp. *eufalcatum* var. *hoffmeisteri*, or it may also be a trinomial when a variety (var.) alone is specified within a species as in the following example.

Anisomeles indica var. *albiflora.*

The terms lesser in rank than form, like race, clone etc. is also in use though to a lesser extent.

The forms are usually developed due to environmental factors and a forma may also be assigned in a species.

For e.g. *Prunus cornuta* forma *villosa.*

The name of the cultivar is not analogous to the botanical variety and according to International Code of Nomenclature for Cultivated Plants (ICNCP). The names are written with a capital letter preceded by abbreviation cv or placed in a single inverted commas. The cultivar name may be used after generic and specific or common names.

For e.g. *Citrullus lanatus* cv. Sugar Baby

The formation of infraspecific epithet follows the same rules as the specific epithet. New cultivar names after 1st January, 1959 are written in the modern language and not in Latin. The only exception to this are names of botanical taxa reduced to cultivar rank.

If the botanical name of a species is changed, the cultivar name remains unchanged. Two or more cultivars in the same cultivar class are not permitted to bear the same name.

Since 1st January, 1959, new cultivar names must not be the same as a botanical or common name of a genus or species. The *Camellia, Eucalyptus,* Rose etc. are not permitted now. The names given previous to this date may of course still be found in use. New cultivar names published after 1 January 1959 require a published or duplicated matter. This may be given in any language. It is recommended that cultivar names be registered with a recognised registration.

Hybrid taxa The principal rank of nothotaxa (hybrid taxa) are nothogenus and notho-species. They are in the same ranks as genus and species, only the terms denoting the ranks differ in order to indicate the hybrid character .

Form genera The species of some fossil plants are based on fragmentary specimens and therefore the genera to which they assigned cannot be named to a family, although they may be referable to a taxon of higher rank. Such genera are known as form-genera *(forma-genera).*

The following are the examples of form genera of fossil plants

1. Form-genera *Dadoxylon* Endl.belongs to *Coniferopsida.,*
2. Form-genera *Pecopteris* (Brongn.) Sternb. *(Pteridopsida),*
3. Form-genera *Stigmaria* Brongn. *(Lepidodendrales),*
4. Form-genera *Spermatites* Miner (seed-bearing plants).

8

TYPE METHOD

Keywords Definition, validity, type concept, type specimen, type variety, type genus, type species, type culture, type locality, type series, nomenclatural types, determination of types, examples, key, significance.

Typification is dealt in chapter 11 of the 2nd section under Articles 7-10 of the Code.

The method of determining the application of names by means of a nomenclatural type is known as the '*type method*' or '*type concept*'.

In order to achieve stabilization of names, the Code has laid great emphasis on typification of taxa. The type concept was first introduced in U.S.A, and the Rochester Code (1892) established the concept of nomenclatural type as a fundamental basis to determine the application of names of taxa. Later, the Vienna Code (1905) ignored the type concept. But again the American Code (1907) confirmed the acceptance of the type concept. In 1918, the Committee on Nomenclature of the Botanical Society of America framed a type-based Code largely based on the type concept. However, the Cambridge Code (1930) reintroduced it as a separate Article, as Article18. According to which "*The application of names of taxonomic groups is determined by means of nomenclatural types....*".

In Recommendations IV of this Article 18 also it is mentioned that, "*when publishing names of new groups, author(s) should clearly mention the type of a new names in a subdivision, the type genus in the family, the type species in a genus, the type variety or specimen in a species...*". The type of family or of any subdivision of a family is the same as that of the generic name on which it is based. For the purpose of citation of a type, the generic name alone is sufficient (Art.10). The type of family is a genus, for instance the genus *Aster* is the type genus for the family *Asteraceae* **(***Compositae***)** **,** the type of a genus is a species, for example the type of the genus *Vernonia* is *Vernonia noveboracensis* (L.) Michx.

Similarly the type of the family *Molluginaceae* is the genus *Mollugo* Linn, which was recently separated from the family *Aizoaceae*. The type of the genus is the *M . pentaphylla* Linn. and this species is typified by one of two specimens (Herm. 3:23 & 4:47), deposited in Hermann's Herbarium, but now it is in the British Museum (Natural History). Its importance lies entirely in fixing the application of names and it was also accepted as a recommendation that, while designating a type specimen for a plant, the place (country and locality) where it is permanently preserved or deposited must be clearly indicated. For example, the type specimen for the species *Vernonia alamanii* DC. is deposited in the de Candolle herbarium in Geneva, Switzerland. However, for the citation of a generic type, it is enough to give the name of the species. There are a number of types such as holotype, isotype, paratype, syntype, lectotype, neotype etc. and their definitions have been introduced in the Stockholm Code (1952) to give stability and validity to the different types. But great stress was laid on the types among the fundamental principles in the Paris Code (1956). The Principle II of the Code states that "*the application of names of taxa of the rank of family or below is determined by means of nomenclatural types*".

The fixing of the type specimen for a species or of the type in general for taxa below the rank of order (family) is considered so important in the Code that Art.35 of the Paris Code (1956), for the first time insisted on the indication of the "type" for validity. Publication, on or after 1st January, 1958 of the name of a new taxon of the rank of an order (family) or below is valid only when the nomenclatural type is indicated.

According to Article 7 of Montreal Code (1961) also, it was mentioned that the application of names of taxa of the rank of family or below is determined by means of nomenclatural types. However, in Leningrad Code (1978) it was further, added that the application of names of taxa above the ranks of a family is also determined by means of the types when the names are ultimately based on generic names (Art.10.7). That is, the principle of automatic typification is extended to those names of taxa above the rank of family that are ultimately based on generic names. For example, the genus *Agave* is the 'type genus' of the family *Agavaceae* and then the order Agavales ultimately based on the name of the type genus *Agave*. This was repeated in the Sydney Code (1983), Berlin Code (1988) and in all other successive Codes, where Articles 7-10 deal with the typification, however in the Berlin Code the determination of types was merged with Article 7 under recommendations, while this was given under a separate heading in the Sydney Code (1983). Typification is "*the process of indicating or designating a type*". "A nomenclatural type (typus) is that 'element' to which the name of the taxon is permanently attached, whether as a correct name or as a synonym (Art.7), or a 'type' that can be defined as an 'element' on which the original validating description of a taxon is based. The nomenclatural type or the

Type specimen is designated as a specimen, indicated by the original author at the time of publication. The nomenclatural type of the name of a species is usually on a herbarium sheet of the specimen or infraspecific taxon is a single specimen or illustration or at times even a photograph or drawing only, except in specified cases (Art.9). For example, small herbaceous plants, and most of the non-vascular plants, the type consists of more than one individual, which was used by the author to give the authentic description and name of the plant, which ought to be conserved permanently must be mounted on a herbarium sheet as they form the record for the future as well as the present reference. The type specimens are of great value and the taxonomists use the type method as a legal device to provide the correct name for a taxon. The type materials are of special importance in nomenclature whenever taxa are subdivided.

Therefore, typification is meant to bring about stability in nomenclature and the types should not change with taxonomic opinions (McNeill, 1981). This means that, when a species is described as new, the author must indicate the type of specimen to which the new species is based upon. That is, a name of a taxon is retained by the type as long as the taxonomic work exist on this earth. Hence the 'type' is merely a nomenclatural concept and has nothing to do with the circumscription of taxa, but help in solving nomenclatural problems.

If a specimen can not be preserved, the type may be a description or figure. The author who first designates a type of a name of a genus or subdivision of a genus must be followed. The type of an autonym is the same as that of the name from which it is derived. For purpose of priority, designation of a type is achieved only by effective publication (Art.29-31), by the typifying author with the clear indication of the type element by direct citation of the term "type" and the type specimens are usually preserved in special folders, cases or valuts.

The name published on or after 1st January 1990, an indication of the type must include one of the words *typus* or *holotypus* or its abbreviation or even its equivalent in a modern language. In case of name of a new species or infraspecific taxon published on or after 1st January 1990, whose type is a specimen or unpublished illustration, and the herbarium or the institution in which the type is conserved must be specified. On or after 1st January 2000, lectotypification or neotypification of a name of a species or infraspecific taxon is valid only when it is indicated by the use of the term lectotypus or neotypus, its abbreviation or its equivalent in a modern language. The specimen selected as a type should belong to a single gathering.

Linnaeus and his contemporaries did not have any idea of proposing a type, when they published the plant names. Therefore typification is a recent concept introduced in plant taxonomy in the year 1930 (Cambridge Code). At the same time all the names were published by these authors without any indication of type cannot be rejected, because they are so numerous and many names are still in use. The Code therefore permits that such names can be validated by subsequent typification.

Likewise, Roxburgh had also failed to preserve the type specimens on which the new names were based, but he made very good illustrations. Several of these illustrations have since been selected as type material. Thus the type material of *Lindernia multiflora* (Roxb.) Mukerjee is Roxburgh's Icones no: 1267 and that of *Limonia crenulata* Roxb. is an example of an illustration in Roxburgh's plants of the Coast of Coromandel. In the sameway, most of the names published by Wallich in his catalogue was also validated by typifying it.

The type of a name of a family or subfamily, not based on a generic name, is the same as that of the corresponding alternate name (Art.18.5& 19.5).The genera that typify the respective alternate specific names should be taken as types of the family names which are not based on generic names, e.g., the genus *Areca* is the type genus of the family *Palmae* (alt. Name-*Arecaceae*) and *Poa* that of *Gramineae* (alt. Name-*Poaceae*) e.g., the genus *Faba* is the type genus of the family *Leguminosae* (alternate name *Fabaceae*), *Aster* that of *Compositae* (alt. Name-*Asteraceae*) and similarly the other four conserved families which do not have ending word 'ceae' also have alternate names, for e.g., *Clusia* for *Guttiferae* (alt. *Clusiaceae*), *Lamium* for *Labiatae* (alt. *Lamiaceae*), *Apium* for *Umbelliferae* (alt.name *Apiaceae*), and *Brassica* for *Cruciferae* (alt.name *Brassicaceae*).

The typification of genera has been a matter of intense debate in the recent past. When we refer to the 'type species' of a genus, whether we are referring to the binomial attributed to the species or the taxon itself should be made clear (Meikle & Brummitt, 1981). However, there is a fundamental difference between the typification of specific names and generic names, in that, the type of the former is objective and may be a specimen, figure or published description, while the latter is more subjectively circumscribed species to which a name itself is applied (Bullock & Hunt, 1966).

The type of a name of taxon assigned to a group with nomenclatural starting point later than 1753 is to be determined in accordance with the indication or description and other matter accompanying its first valid publication. The term 'Type culture' is a culture derived from a holotype, lectotype, or neotype. Similarly the 'Type genus' the genus, that is the type of the family and 'Type locality' is the exact geographical site at which the type of a species or subspecies was collected. All the specimens on which the original description of a species or subspecies was based, called as the 'type series' or 'type material'.

Nomenclatural type There are different kinds of nomenclatural types introduced by the Stockholm Code (1952). These are holotype, isotype, paratype, syntype, lectotype, neotype and so on.

1. Holotype A holotype is a specimen or other element designated by the original author as the nomenclatural type i.e. the element to which the name of the taxon is permanently attached.
2. Isotype It is a duplicate specimen of the holotype. At the time of collection four or five plant specimens are collected and pressed. At the time of designation, the author select one of the specimens as the holotype, and the rest are considered as duplicates, called the isotypes.
3. Paratype A paratype is a specimen cited in the protologue with the original description other than the holotype. But the specimens are collected at different times and will have different field number.
4. Syntype A syntype is any two or more specimens or elements used by an author when no holotype was designated, or when any one of two or more specimens was designated simultaneously as the type.
5. Lectotype A lectotype is a specimen or other element selected from the original material to serve as the nomenclatural type, when the holotype was not designated at the time of publication, or when the designated holotype is missing or lost.
6. Neotype A neotype is a specimen or an illustration selected to serve as the nomenclatural type as long as all of the material on which the name of the taxon was based is missing.
7. Topotype A topotype is a specimen collected at the same place as was the type, i.e. from the type locality.
8. Cotype A cotype is a second specimen from the same plant from which the holotype was collected.

9. Autotype When a name is automatically established for a taxon, merely as a consequence of the proposal of another taxon and the specimen designated in such case is called an 'autotype' or autonym. That is the one which includes the type specimen of the taxon must have the same epithet as that of the originally described species. For e.g., when the species *Rosa gallica* is divided into variety, it becomes *Rosa gallica* var. *gallica,* which includes its type and the other species is designated as *Rosa gallica* var. eriostyla by Keller. Similarly *Acacia nilotica* ssp. *nilotica* and *Crotalaria willdenowia* DC subsp. *willdenowian* are the examples of autonym.
10. Epitype A specimen or an illustration selected to serve as an interpretative type, when the holotype, lectotype or previously designated neotype, or all original material associated with a validly published name, is demonstrably ambiguous and cannot be critically identified for purposes of the precise application of the name of a taxon, when an epitype is designated, the holotype, lectotype or neotype, that the epitype supports must be explicitly cited (Art.9.7).
11. Schizotype Any fragment of a type is known schizotype.
12. Paralectotype One more type, called paralectotype, was proposed by Hansen & Sieberg (1984) for specimens which were left after the selection of lectotype out of the syntypes.
13. Homeotype It is a 'specimen compared by a competent observer with the holotype, lectotype or other primary type of the species' (Frizzell (1933).

The following are other substitute types of minor importance used in the Code:

1. Interpretotype: It is that constituent element not clearly proved to have served as a basis for a

description, but designated by the author of a taxon by the name in question.

2. Merotype: It is a fragment of the original holotype which has been divided into two or more pieces after having been used as a basis for description.
3. Synonymotype: It is that constituent element of a taxon quoted by the author as being identical with his newly described taxon.
4. Clastotype: It means the portion of the same clone.
5. Ideotype and metatype: They are specimens examined by the nomenclator himself and considered by him to be in complete accordance with the nomenclatural type, and partially correspond to each other.
6. Icotype: An Icotype, often considered to be synonymous with an Iconotype (Frizzell, 1933 is a photograph or drawing of a type) is a specimen which confirms to the characters of the recognised type of the taxon concerned.

Note The term type used alone is invalid, and it generally refers to the holotype.

Even though, theoretically, we still have type species, type genera etc., in practice there is only one kind of type exist, that is the species. The Code also strictly recommend a holotype should be designated and deposited in a well established herbarium or institution and conserved permanently.

Determination of different types The procedure for the determination of different types given by the Code are as follows:

1. **Holotype** A holotype is a specimen or other element used by the author for describing a new species as the

nomenclature type (Art.7.3).The holotype is selected by the original author from a single gathering made by him at one time, and mentioned properly at the time of the original publication.

As long as holotype is extant, it automatically fixes the application of the name concerned. The type of a name of a species is usually a single specimen, whereas for most of the non-vascular plants and small herbaceous plants, the type may consist of more than one individual fixed permanently on one herbarium sheet. Once the holotype is designated by the original author it must not be changed or rejected and any type chosen after the original publication can not be regarded as the holotype. If the original author has not mentioned the holotype, at the time of his publication, a lectotype or neotype can be selected by later authors. It is to be noted that neither the original author of a taxon, nor any one else can choose a holotype at any time after the publication of the taxon; however such later choice might always establish either lecto or neotypes.

The Code recommend that publication on or after 1.1.1958 of the name of a new taxon of the rank of family or below is valid only when the nomenclature type is specified and it should be deposited in a well established Herbarium.

For example, Chandrabose in his original publication of a new species of *Polygala* clearly indicates the type as given below:

1. *Polygala jacobii* Chandr. in Bull.Bot.Surv.India 9 (1-4): 288-289. 1967. Holotypus, Chandrabose 28777A (MH).

   (MH stands for Madras Herbarium and later on MH was shifted to Botanical Survey of India, Southern Circle in Coimbatore).

2. *Phyllanthus mukerjeeanus* Mitra et Bennet in Bull. Botan. Soc. Bengal 19(2):145-146.1966. Holotype Bennet 621A (CAL).

2. **Isotype** Any duplicate specimen of the holotype, a part of a single collection made by a collector at the same place and time is known as isotype.

For example if a new species is based on a single collection, having five specimens and five herbarium sheets made out of it bearing the same field number, the original author designates one herbarium sheet among them as a holotype and the rest of the four specimens as isotypes. Therefore, the duplicate specimens of the holotype are called isotype.

For example Chandrabose after designating one specimen as holotype from the collection of *Polygala jacobii*, he considered the rest of it as isotype.

a. *Polygala jacobii* Chandr. In Bull. Bot. Surv.India 9(1-4). Isotypus Chandrabose 28777B-E (MH).

b. *Phyllanthus mukerjeeanus* Mitra et Bennet in Bull. Botan. Soc. Bengal 19(2): 145-146, 1966. Isotype Bennet 621 B-C (CAL).

3. **Syntype** When no holotype is designated by the original author at the time of publication, all the specimens cited by him are known as syntype.

For example, when *Cynodon barberi* a new species of a grass from South India was described by Rangachari & Tadulingam (1916),failed to designate one specimen as holotype. Therefore all the specimens cited by them in their original publication became syntype.

4. **Paratype** A specimen cited by an author with the original description other than the holotype or isotype at the time of describing a new taxon is known as paratype.

If a new taxon is described based on more than one gathering made at various time and seasons, the author would designate one collection as holotype and isotype. The rest of the collections made either from the same localities or from different localities by the same or different authors with different field numbers will constitute paratype.

For instance, if the new taxon is collected with flowers in one season and with fruits in another season, these two gatherings will bear two different field numbers. The author will select one collection for designating holotype and isotype, and the next gathering forms the paratype.

For e.g., *Polygala jacobii* Chandr. In Bull. Bot. Surv. India 9 (1-4): 288-289. 1967. Paratypus Rangachari s.n. (Acc. no.2444, MH).

5. **Lectotype** A lectotype is a specimen chosen by a later worker from the original materials to serve as a nomenclatural type, when no holotype was designated at the time of publication by the original author or when the holotype was missed or lost or destroyed. The specimen chosen from available material exactly matches with the description of the original author is selected to serve as a lectotype (Art.7.5).

   The duplicates of lectotypes become isolectotype.

In choosing a lectotype, all aspects of the protologue should be considered as a basic guide. If there is no holotype, a lectotype can be selected from other types such as isotype or paratype or even from syntype if available.

For e.g., Rangachari and Tadulingam (1916) while describing a new species (J. Bombay Nat. Hist. Soc.24:846,1916), *Cynodon barberi* from South India, cited many specimens in the protologue, but they failed to indicate one specimen as the holotype at the time of original publication of the species. Hence all the specimens cited in the original publication became syntype.

Jain (Bull. Bot. Surv. India 8:204.1967) after revising the original description of *Cynodon* and the herbarium sheets prepared by Rangachari and Tadulingam selected one of the specimens –"Madras Herbarium, South Indian Flora No.13715 deposited in CAL-as the lectotype and it was deposited in MH(Madras Herbarium, now functioning at Botanical Survey of India, Southern Circle, Coimbatore), to serve as nomenclatural type.

A lectotype always has precedence over a neotype, because the details of lectotype were already studied by the original author.

6. **Neotype** A neotype is a specimen or other element selected to serve as nomenclatural type only when all the originally described materials or duplicates on which the name of a taxon based are extant (Art.9.11).

The early botanists did not indicate any type specimen as is done today. To these early botanists, the plant species were based upon all specimens, illustrations and descriptions within the limits of the species. These materials or elements or everything associated with the name at first publication are collectively known as the protologue.

Therefore, in selecting a neotype more care and critical knowledge are essential as the reviewer usually has no guide to refer and verify, except his own judgement as to what plant material best fits the protologue.

For e.g., when Blatter and Hallberg described the new species *Farsetia macrantha* (J.Bombay Nat. Hist. Soc.26.220,1918) three numbers such as Blatter 5785, 7300 and 7305 was cited in the protologue. These specimens were deposited in the Blatter Herbarium (BLAT), Bombay, where the entire collection of Blatter, including the 'holotype' have been preserved.

After a critical study on this species Bhandari (J. Bombay Nat.Hist. Soc.72:606, 1975) could not trace the type specimen (holotype) in the Blatter Herbarium and assumed that all the original material of *Farsetia macrantha* has been lost or destroyed. Hence, he selected Bhandari 507, and designated it as a neotype, "exemplifying the true application of the name *Farsetia macrantha* Blatt & Hallb. The specimen comes from the type locality (place of collection) and is in complete agreement with the original diagnosis of the species".

Neotype is otherwise called proxytype.

Material cited and/or seen by the author or duplicates thereof is known to exist:

Types designated by the author:

A single collection cited by the author as type:

The one specimen designated by the author as type... Holotype

Duplicates thereof..Isotype

Several collections cited by the author besides the holo- and isotypes:

Specimens seen by the author.................... Paratype
Duplicates thereof................................ Isoparatype

Types not designated by the author:

Several collection cited by the author

Collectively as types....................................Syntype
Duplicates thereof.. Isosyntype

Types designated by a subsequent author(s):

Materials selected from or cited by the original author..... Lectotype

Remaining syntypes after designation of one of them as lectotype...Paralectotype

Material cited and /or seen by the author and duplicates thereof, are not known to exist:

Subsequent author designates a new type......Neotype
Duplicates thereof.. Isoneotype

Key A key is provided to identify the different nomenclatural types introduced by the Code. Key is a device or tool useful to identify an unknown taxon.

The aim of the key is to provide an ease and certainty of identification of any taxon (Davis and Heywood, 1963).

The following key (modified from Faegri, 1983) is useful in recognising the different nomenclatural types.

Guide for determination of types The guidelines for the selection of the nomenclatural types by any author(s) are as follows:

1. A holotype designated by the original author at the time of publication of the taxon is a specimen and his choice is final.
   - ★ As long as a holotype is extant it fixes the application of the name of the taxon concerned and it can not be changed.
   - ★ At the time of publication, if an author indicated only one specimen, that element must be accepted as a holotype.
   - ★ It is to be noted that no one is allowed to choose a holotype after the publication of the taxon; however any later choice if made always establishes only a lectotype.
2. In most of the cases in which no holotype is designated there will also be no paratypes, since all the specimens cited by the author will become syntypes.
3. A nomenclatural type published for a new name or epithet as a substitute (nomen novum) was the same type of the older name.
4. An isotype is any duplicate of the holotype and it is always a specimen.

5. A lectotype is selected from the original material on which a taxon was based; that is a lectotype must be chosen from the materials actually studied by the original author for his publication of the name of the taxon and included in the protologue.

   - ★ If the holotype designated by the original author has been lost or destroyed, an isotype (duplicate of holotype) if specified by the original author is existing, can be selected as a lectotype.
   - ★ If the holotype was not indicated by the original author at the time of publication of a taxon, and if syntypes exist, one of them is chosen as a lectotype.
   - ★ If neither an isotype nor a syntype is available, a paratype if it exists, may be chosen as the lectotype.
   - ★ The first choice of a lectotype must be retained unless the original material is rediscovered, or the choice made was proved incorrect.

6. A neotype may be selected only when a holotype or a previously designated lectotype has been lost or destroyed and all other original materials cited by the author are believed to be lost or destroyed or all other original available materials differ taxonomically from the destroyed type. (Art.9.12).

   - ★ In selecting a neotype, one has to be very careful and must have critical knowledge as the reviewer has no guide to refer and verify, except his own judgement as to what best fits the protologue.

7. In case the type of a name of a species or infraspecific taxon of non-fossil plants cannot be preserved as a specimen or if such a name is without a type specimen, the type may be represented by an illustration.

8. The type of fossil plants of the rank of species or below is the specimen and its figure either accompanied or

cited in its valid publication of the name. If figures of more than one specimen are given or cited one of these specimens must be chosen as the type.

9. The type (holo-, lecto-, or neotype) for a species or its lower rank is a single specimen mounted on a herbarium sheet except for small herbaceous or non- vascular plants more than one specimen are mounted (Art.9). Type for the genus or its subdivision is the type of species for a family and its subdivision is the same as that for the genus (Art.10).

10. The type specimen must be preserved permanently and can not be a living plant or a culture.

11. For establishing a taxon of fossil plants, one whole specimen is considered as nomenclatural type and if the type material of a name of a fossil plant is available in cut pieces, all parts used for description or diagnosis should be clearly marked.

12. When a holotype or lectotype or a neotype is represented by an illustration,that specimen should clearly help to determine the application of the name and the collected data of the illustrated material also should be given for a new taxon.

13. The author who first designates a lectotype or a neotype must be followed but his choice of selection can be superseded by the holotype or in case of a neotype or any of the other original material is rediscovered.

14. Lectotypification or neotypification of a name of a species or infraspecific taxon by a specimen or unpublished illustration becomes valid on or after 1.1.1990, only when the herbarium or institution where the type is conserved must be specified.

15. In case of selection of lectotype needed for the name of a fossil species, it should be a specimen illustrated at the time of the valid publication of the name.

Significance of type specimen The type method helps in solving a number of taxonomic problems. In this respect, the nomenclatural types are considered as the 'centroids', around which we develop our concepts of taxa and to which the names are attached. The question is, how far one can go from the centroid in drawing the circumference of the taxa, which a person(s) is dealing with, to find out the correct position of the taxon or taxa in question. In such disputes the type material of the plant could be helpful in many ways.

The following are some of the examples which explain the working of the type method in tackling taxonomic problems:

a. In detecting new taxa:

i. The following species were included in the genus *Pinus* when it was first published by Linnaeus in 1753.

Pinus sylvestris

P. pinea

P. taeda

P. cembra

P. strobus

P. cedrus

P. larix

P. picea

P. balsamea and *P. abies*.

Later on, when the species of the *Pinus* was analysed, it was understood that, Linnaeus's concept of the genus *Pinus* was far too wide. Therefore, his species were categorized based on their similarities and dissimilarities into five distinct genera and are as follows:

Genus-1 *Pinus cedrus*

Genus- 2 *Pinus larix*

Genus-3 *Pinus picea* & *Pinus balsamea*

Genus-4 *Pinus abies*

Genus-5 *Pinus sylvestris, P. pinea, P.cembra, P.strobus* & *P. taeda*.

After classifying the species, the question raised is, 'To which of these five genera is the actual name *Pinus* to be applied? To solve this problem, one has to use the type method in such a way that, the designated type specimen by the original author is compared with the categorised five genera. The one (category) in which the type of the name *Pinus* matches actually constitutes the genus *Pinus*. The type of a generic name is a species and in the case of the name *Pinus*, it is the species *Pinus sylvestris*. This species is placed in the 5th genus (*P. sylvestris, P. pinea, P.cembra, P.strobus* & *P.taeda*) and it is to this genus that the name *Pinus* to be applied. The other four genera (1-4) must be removed from the name *Pinus* and they should be kept under different genera such as *Cedrus, Larix, Picea* and *Abies*.

ii. The genus *Thottea* Rottb. (=*Apama* Lam.) belongs to the family *Aristolochiaceae*, which distributed in India, was thought to have only a single species *T.* ***siliquosa*** (Lam.) Ding Hou, with uniseriate stamens united in three bundles. However, the specimen collected from the Western Ghats along the Western Coast of India revealed that there are at least three different categories involved in it, which differ from each other in their floral characters. Now, the question is which one of these three plant materials is the actual *T. Siliquosa* ?

To get the answer to this question, one had to compare the collected specimens with the original type material of the species (Rheede, Hort. Malab.6:51.t.28.1686). The compared materials revealed that the widespread taxon with pendent, small, purple flowers with deeply three-partite perianth is revealed as *T. siliquosa*. Hence, the other two categories have been described as new taxa , viz. *T. duchartrei* Sivar.*et al.*, Ind. J. Forestry 8: 265-8.(1985) and *T.ponmudiana* Sivar (Pl.Syst. Evol.150:201-4:1985).

iii. Similarly, the species *Heliotropium indicum* Linn., which is commonly occurring is observed to consists of two different elements, which is quite distinct from each other. To differentiate between them, it is very essential to findout which one of them is *Heliotropium indicum* proper. To solve this problem, the type specimen is used for comparison. The plant which resembles with the type material based on which Linnaeus described, should be attributed to the Linnaean binomial. The other plant material is therefore described as a new species under the genus *Heliotropium* as *H. keralense* Manilal et Sivarajan (1972).

iv. When the species *Portulaca pilosa* is divided into two subspecies, the one which includes the type specimen of the species automatically gets the name subsp. pilosa (autonym) as *Portulaca pilosa* subsp. *pilosa* and for the other, a new subsp. name grandiflora is given. Therefore, it is cited as *Portulaca pilosa* subsp.*grandiflora*.

a. In the Sub-division of taxa:

i. It is the type method which decides the correct name when the species are subdivided. For example the application of names when one plant is divided into two or more or when the circumscription of a plant is widened so as to include two or more already named taxa.

The specimen with which the original type resembles, inherits the original name or epithet. Thus, the family *Aizoaceae* when divided, the part in which the type genus *Aizoon* was included retained the original name, while the rest was segregated into a new family as *Molluginaceae* with *Mollugo* L. as the type genus.

ii. Plant name published without designating type is considered as illegitimate and has to be rejected. *Rotala malampuzhensis* R.V. Nair (1977) is an e.g. of an illegitimate name, because R.V. Nair failed to indicate the type specimen for this species at the time of publication. But later the name was validated by Cook (1979) by lectotypifying it.

c. To detect wrong identifications:

i. Linnaeus's son (Linn.f.) described a species *Spermacoce stricta* based on a Rumphian element. However, the type material of this plant was later studied by Merrill (1917) and Bremekamp (1959) and both of them concluded that, the studied material is a *Hedyotis* and not a *Spermacoce*. Therefore, the specific epithet *stricta* must not be retained in the genus *Spermacoce*, due to wrong identification and the common Indian material so far known under the name *S.stricta* has now been given with a new specific epithet and hence written as *S. ramanii* Sivar & Nair (1986).

ii. The species *Borreria rosea* (Sivar. & Mani) Sivar. 1982, was later on observed to be conspecific with *Spermacoce pusilla* (Wall.) DC. (1828), and *Borreria eradii* Ravi (1970) has been reduced to *Spermacoce latifolia* Aubl. (1979). Thus, type material helps to detect the wrongly identified taxa.

iii. All the collected materials of the genus *Spilanthus* Jacq. from India has been assigned to a single species, *Spilanthus acmella* Murr. After revision, Koster and Philipson (1950) concluded that, the type specimen of the species is *Blainvillea* and not *Spilanthus*. Therefore, according to the rules, the specific epithet *acmella* must not be retained in the genus *Spilanthus*. As a result, all these specimens are included in *S. paniculata* Wall ex DC. Further studies also revealed that there are at least six species of this genus available in India.

d **In redefining taxa** Until recently the species *Mollugo pentaphylla* L. was considered to be a highly polymorphic species that also included *Mollugo stricta* Linn. Linnaeus recognised two species based on their foliar and inflorescence characters and subsequent authors also considered these species as intergrading.

But on revision of the circumscription of these two taxa, it was concluded that, they are of two different species, differing from each other not only in their aforesaid features, but also in their seed coat patterns and sepal venation patterns. Therefore, the two were separated from each other and designated as *Spermacoce hispida* Linn. and *Spermacoce articularis* Linn.f. In addition the circumscription of these taxa has also been altered due to similar studies.

e. **To detect homonyms** When the same name is assigned to different taxa, these are termed homonyms and are illegitimate in their later application. So, illegitimate homonyms have to be rejected according to ICBN. The question here is, how can one detect such homonyms? Nomenclatural types hold the key to these problems.

i. The authors Bennet & Raizada (1981) while studying the genus *Courtoisia* Nees (1834) of *Cyperaceae* found that, it is a later homonym of *Courtoisia* Marchand (1830), a genus of Lichens. Since the lichen member was published earlier, the authors Bennet & Raizada retained the original name of the genus *Courtoisia* as such for lichen and they have given a new name to the genus of the family *Cyperaceae* as *Indocourtoisia*, instead of having it as *Courtoisia* which was published later in 1834. With such widely different plant species one may easily detect the homonym from the descriptions.

ii. Whereas in the case of closely related taxa, one has to obtain the help from the nomenclatural type of the plant concerned. For example the genus *Tainopsis* Hayata (1914) of the family *Orchidaceae* was described based on a Formosan species. However, the same name was given to a distinct monotypic genus of Indian orchid by Schlechter (1915). According to ICBN, the name proposed by Schlechter forms the later homonym and it is to be rejected. But Bennet & Raizada (1981), after analysing the type specimens of both the plants, have identified this error and therefore they have validated Schlechter's plant by giving a new name, *Neotainopsis*.

9

PRINCIPLE OF PRIORITY AND ITS LIMITATIONS

Keywords Operative principle, rule of priority, recommendations, limitations , applications, examples, history of Triticum sp., conserved species, examples.

Introduction One of the important operative principles of ICBN is the "principle of priority". Out of six principles of ICBN, the III and IV deal with the principle of priority (Art.11-15). According to principle III, "*the nomenclature of a taxon is based upon priority of publication*" and IV states that " *Each taxonomic group with a particular circumscription, position and rank can bear only one correct name, the earliest that is in accordance with the rules*". The principle of priority is therefore concerned with the selection of a single correct name for a taxonomic group. It is the most important principle of nomenclature has served to provide a stable method of naming taxonomic groups. This method involves identification of legitimate and illegitimate names and rejecting the latter, and selecting a correct name from among the legitimate ones.

Since strict adherence to the principle of priority was also recommended in the Rochester Code (1892), the huge collection of plant specimens that were already made practically from all over the world had to be identified and named properly. Hence, the identification was done by the many botanists, but quite independently. As a result, the widely distributed plant species of different countries received different names at the hands of these botanists. This has ultimately resulted in naming of same species with different names in different parts of the world and also resulted in the publication of many names for plants, that exceeds the actual number of taxa available in the world flora. For example, the Indian Coral tree (*Erythrina variegata* L.) has about as many as 200 scientific names (Krukoff,1972). Similarly many ornamental plants, medicinal plants and economically important grasses have numerous names and most of which are confusing.

Thus it clearly indicates that several names have frequently been attributed to one and the same taxon. To avoid such confusion, it becomes very essential to eliminate the duplication of names and also one has to select a valid name out of the several names published for a taxon. The Code has certain provisions to rectify such problems by applying the rule of priority. This is called "The Principle of Priority" (Art.11-15). Article 11 of the Code insists that each family or taxon of lower rank with particular circumscription, position and rank can bear only one correct name, the earliest name validly published for the taxon, except in the cases of the conserved nine families, and one subfamily listed below for which alternate names are permitted

Conserved family name	Alternate name
Cruciferae	Brassicaceae
Umbelliferae	Apiaceae
Leguminosae	Fabaceae
Guttiferae	Cluciaceae

Compositae	Asteraceae
Labiatae	Lamiaceae
Palmae	Arecaceae
Gramineae	Poaceae

and the subfamily is a *Papilionoideae.*

In case more than one legitimate name exists for a taxon, the correct name is the earliest legitimate name available in the same rank, except for nine families as named earlier for which alternate names have been given. At the family and genus levels, exceptions are also permitted where names are conserved. For the taxa at the species level and below, the correct name is either the earliest legitimate name or a combination based on the earliest legitimate basionym, unless the combination becomes a tautonym or later homonym, rendering it illegitimate. The principle of priority is not mandatory for names of taxa above the rank of a family; however, it is recommended by the Code (Recommendation 16B) that while selecting from among typified names for such a taxon, the author should follow the principle of priority.

For example, the common apple was first described by Linnaeus (1753) under the name *Pyrus malus*.The species was subsequently transferred to the genus *Malus*, but the combination *Malus malus* (Linn) Brit. (1888) could not be taken as a correct name because it becomes a tautonym. The two other binomials under *Malus* available for apple include *M.pumila* Mill.(1768) and *M. domestica* Borkh.(1803), and thus *M. pumila* Mill. (1768), being earlier of these two is selected as the correct name for apple and the citation written as:

Malus pumila Mill. (1768)

Pyrus malus Linn. (1753)

M. domestica Borkh. (1803)

M. malus (Linn.) Britt. (1888) – tautonym.

The rule of priority states: "*For any taxon from family to genus inclusive, the correct name is the earliest legitimate one with the same rank except in cases of limitation of priority by conservation*". Conservation of name means, some names such as family, genera and species, even though they are not the oldest legitimate names, but are used in preference to the earlier names.

The Code also states that "*The principle of priority does not apply to names of taxa above the rank of family* ". However, it is recommended that while choosing among typified names for such a taxon, the author should follow the principle of priority.

A name of a taxon has no status in this Code unless it is validly published (Art.12). The rule of priority provided opportunity to many taxonomists to unearth earlier or earliest names and use them instead of current names. This condition created much confusion instead of bringing the names into order. Therefore, application of principle of priority finds many problems. As a result this rule had to be modified, and several changes were made to avoid disadvantageous changes in the nomenclature of certain genera, families and intermediate taxa, for which Code provides list of names that are conserved in preference to the earlier published names (nom.cons.), earlier legitimate names rejected (nom.rejicienda) and given in Appendix II and III. by resolution of the IBCs (Art.15).

One such example is *Desmodium* Desv. a leguminous plant, which was published in the year 1813. But plants of this genus were also included under two other genera, namely *Meibomia* Heist ex Fab. (1759) and *Pleurolobus* J.St.-Hil. (1812). All the three genera have the same circumscription.

According to the principle of priority, *Meibomia* should have been selected as a correct name but since *Desmodium* is better known and widely used, it has been conserved and accepted as a correct name than the earlier published names.

The genus *Tectona* L.f. (1781) is conserved against its earlier name *Theka* Adans. (1763) due to its very long usage in botanical publications; the obsolete generic name *Theka,* though earlier, is rejected.

Similarly, the family name *Sterculiaceae* Lindl. 1830 is considered as a conserved name although not published earlier. But the earlier published name for the above family is *Byttneriaceae* R. Br. 1814 (Lawrence, 1951) which is to be rejected.

A list of only conserved family and generic names are given by the earlier Codes, but now a list of conservation of specific names was also authorized and is restricted to names of species of major economic importance (Greuter, 1981) and was adopted at the 13th and 14th International Botanical Congress held at Sydney, Australia in 1981 and Berlin (1987) and the first species conserved by the Berlin code (1988) is *Triticum aestivum* Linn.

The Code also proposed certain limitations for principle of priority, with some important dates considered to be the beginning for different plant groups. The valid publication of names of plants of different groups is treated as beginning or starting at some prescribed dates corresponding to the publication of some works (Article-13). The botanists have agreed with the opinion that, the priority of nomenclature (starting date) for vascular plants is applicable from May 1st, 1753 which is the date of publication of Carolus Linnaeus' Species Plantarum.

For example for vascular plants Spermatophyta and Pteridophyta, 1st May, 1753, (Species Plantarum) is considered to be the beginning or starting date. It means that, one can unearth the earliest name only as far back as 1753 whereas the names published prior to this date need not be considered in applying the rule of priority and those names will not be taken into consideration. That is the name published for Spermatophyta before this date is ignored while deciding priority.

The following are the examples that illustrate the application of principle of priority:

i. Thus there should be only one valid name for a particular plant, but it may have many synonyms. For the above date of publication has to be mentioned for each name.

ii. This rule is not applicable to the names of taxa above the rank of family.

iii. Any association of a specific epithet with a generic name is known as a combination. All those names presented in accordance with the rules of nomenclature are termed legitimate and the names contrary to the rules of nomenclature are termed illegitimate.

1. According to the rule of priority, when two or more taxa of the same rank are united, in general it is by the oldest or the earliest legitimate name or epithet should properly be selected. By the oldest means the validly published first name. The name by which a taxon is properly known is called its correct name by the Code. For example, when K. Schumann united three genera such as

   Sloanea L. (1753)

   Echinocarpus Blume (1825) and

   Phoenicosperma Miq (1865) - rightly selected the oldest or the earliest of these three generic names, *Sloanea* L

(1753) for the combined genus and the other two names would be the taxonomic synonyms must be rejected.

2. The name *Cassia surattensis* was given to a taxon by Burmann (Flora indica) in 1768.The same taxon was named by Lamark as *Cassia glauca* in 1785. When the principle of priority is applied to this situation, Burmann's name (*Cassia surattensis*) gets the priority.
3. According to the rule of priority, the names *Trianthema portulacastrum* L. (1753) and *Fagraea ceilanica* Thumb. (1782) are the correct names, but these taxa are also known under the binomials *Trianthema monogyna* L. (1767) and *Fagraea obovata* Wall. (1824) respectively are to be rejected.
4. Similarly, the name *Lawsonia inermis* L. (1753) has priority over *Lawsonia alba* Lamk. (1789).
5. Likewise, *Mollugo pentaphylla* L. (1753) has priority over *M.stricta* L.(1762), and *Foeniculum vulgare* Gaertn. (1768) has priority over *F. capillaceum* Gillb.(1781).
6. The name *Xyris* was published by Carolus Linnaeus in his Species Plantarum (1753). The same plant was also named as *Schismaxow* and published by Steudel in Bot. Zeitung 14, 1856. So the priority should be given to *Xyris* as it was published earlier than *Schismaxow.*
7. The name *Mondo* was published by Adanson in 1763. The same plant was also named as *Ophiopogon* by Kerl Gawl in 1807.So the name *Mondo* was retained as a valid name due to its priority.
8. The nomenclature history of *Polygala* shows that, this plant was given with the following four names by different authors in different dates. These are, *Polygala* Linn. (1753), *Poligala* Neck. (1768), *Polygaloides* Agosti. (1770), and *Phylax* Nor. (1770).

According to the rule of priority, the earliest name *Polygala* Linn. (1753) is the correct name for the combined genus and the other names are the synonyms, must be rejected.

9. The three commonly known binomials for the same species of *Nymphaea* are *N.nouchali* Burm. f.,1768, *N.pubescence* Willd., 1799 and *N. torus* Hook.f. et T., 1872. According to the principle of priority *N. nouchali* Burm.f. is selected as the correct name as it was published earlier and the other two species are regarded as synonyms. The citation is written as :

   Nymphaea nouchali Burm.f., 1768

   N. pubescence Willd. 1799

   N. torus Hook.f. et T., 1872.

10. The two species of Almond, *Amygdalus communis* Linn. (1753) and *A. dulcis* Miller (1768) are regarded as synonymous.

When the species *Amygdalus communis* Linn. (1753) was united with the genus *Prunus* by Archangeli in 1882, the combination becomes *Prunus communis* (Linn.) Archangeli (1882). Later, it was discovered by Webb that, the binomial *Prunus communis* had already been used by Hudson in 1762 for a different species. Hence, *P. communis* (Linn.) Archangeli (1882) became a later homonym, which was rejected. Therefore, Webb used the next available epithet (basionym) under *Amygdalus dulcis* Mill. 1768 and formed the combination as *Prunus dulcis* (Mill) Webb.1967 as the correct name for Almond. Therefore, the binomial *Prunus amygdalus* Batsch.(1801), could not be used due to the presence of earlier epithets.

The citation for Almond is :

Prunus dulcis (Mill.) Webb. 1967

Amygdalus dulcis Mill. 1768 –basionym

A. communis Linn. 1753

Prunus communis (Linn.) Arch. 1882 (non. Huds, 1762)

P. amygdalus Batsch, 1801.

11. The Code does not give importance to priority of publication, when both names of a taxon were published on the same date. For example the generic names *Thea* L. Species Plantarum . 515 (May,1753) and *Camellia* L. Species Plantarum 698(August,1753) are treated as having been published simultaneously in May 1753. Sweet in 1818, who was the first person to unite the two genera, chose *Camellia* for the combined genus, citing *Thea* as a synonym and his choice should be strictly followed.
12. Similarly, Linnaeus published *Waltheria indica* and *Waltheria americana* in Species Plantarum 673. 1753 (i.e. on the same date -1st May, 1753). Robert Brown, when first united the two species, selected the name *Waltheria indica* L. for the combined species, and this name should be retained.

Limitations of priority (Arts. 13-15) The application of principle of priority has the following limitations:

1. **Specified starting date** It can not be applied to all names published. Instead, valid publication of names of different groups of plants is treated as beginning on various dates as provided in the code. The rule of priority starts with the date of publication of *Species Plantarum* by Linnaeus in 1st May, 1753.The different starting dates for the valid publication of different plant groups as provided in the Code (Berlin Code, 1988.Art. 13) are:
    1. May 1st, 1753 (Linnaeus, Sp. Pl. ed. 1) – Spermatophyta and Pteridophyta.
    2. Jan.1st , 1801 (Hedwig, Sp. Muscorum) – Mosses except Sphagnaceae.

3. May 1st, 1753 (Linnaeus, Sp. Pl. ed. 1) – Sphagnaceae and Hepaticae.
4. May 1st, 1753 (Linnaeus, Sp. Pl. ed. 1) – Fungi including Myxomycetes and Lichen fungi.

Dec.31st, 1801 (Persoon, Synopsis Methodical Fungorum) -Uredinales, Ustilaginales,Gastromycetes.

Jan.1st, 1821 (Fries, System Mycologicum, vols. 1-3 and Elenchus, Fungorum., vols.1-2) – Fungi Caeteri except Myxomycetes and Lichen forming fungi.

5. May 1st, 1753 (Linnaeus, Sp.Pl.ed. 1) – Algae except the following :

Jan. 1st, 1892 (Gomont, Monographie des Oscillaries, Ann.Sci. Bot. Ser. 7, 15: 263-368; 1619) – ostocaceae – Heterocysteae.

Jan. 1st, 1848 (Ralfs, British Desmidieae) – Desmidiaceae.

Jan 1st, 1900 (Hirn, Monographie and Iconographie der Oedogoniaceen,Acta Soc. Sci. Fenn.27(1)) – Oedogoniaceae.

Dec.31st 1900 (Sternberg, flora der Verwelt Schlotheim Petrefactenkunde) –All Fossil groups.

The two volumes of '*Species Plantarum*' May, 1753 and August, 1753 are to be treated as published on May, 1st, 1753.

The plant names published before the above said dates for respective groups need not be considered in applying the rule of priority.

2. **Application upto family level** The principle of priority is not applicable to names of taxa above the rank of family but is applicable only to the rank of family and below.
3. **Not outside its own rank** In choosing a correct name for a taxon, names or epithets available at that same rank are to be considered. A name or an epithet will have no priority outside its own rank.

For example, when the rank of a genus or infrageneric taxon is changed, the correct name is the earliest legitimate available in the new rank.

Cyperus iria var. *rectangularis* Kukenth (1935) when raised to the specific rank by Bennet, is correctly called *Cyperus alulatus* Kern (1952), and not *Cyperus rectangularis* (Kukenth.) Bennet (1969). In no case does a name or an epithet will have priority outside its own rank (Art.60). The varietal epithet *–rectangularis* though published earlier, has no priority outside its own rank and hence the combination *Cyperus rectangularis* (Kukenth.) Bennet (1969) is illegitimate and must be rejected.

When a correct name at the same rank is not available, a combination can be made using the epithet from another rank.

4. **Nomina conservanda (Nom. cons.)** The principle of priority cannot be applied in the case of conserved names. The strict application of rule of priority has resulted in numerous name changes. In order to avoid the name changes of well- known families and genera, especially those containing many species, a list of conserved family and generic names (Nomina familia and nomina generica conservanda et rejicienda-Appendix IIIA) has been prepared and published in the Code with relevant changes. Such names are listed as *nomina conservanda* and are to be used as correct name replacing the earlier legitimate names, which are to be rejected and constitute nomina rejicienda (nom. rejic.).

For example, the family name *Theaceae* D. Don, 1825 was thus conserved against its earlier name *Ternstroemiaceae* Mirbe in 1813. Thus the name *Butea* Willd (1802) was conserved against its earlier name *Plaso* Adans.(1763).The same way, *Cajanus* DC (1813) was conserved against its earlier name *Cajan* Adans.(1763). Similarly the genus *Sesbania* Scop.,

1777 was conserved against its earlier names *Sesban* Adans. 1763 and *Agati* Adans.1763. The list of such name is open for discussion. Before conserving any such name, the matter has to be exhaustively discussed in the pages of Taxon, the official organ of the International Association for Plant Taxonomy and put to a vote. The Botanists are authorised to use such conserved names after the approval by the General Committee pending for final approval by the subsequent International Congress. Such conservation puts an end to changes in generic names due to nomenclatural conventions such as the law of priority does not rule out changes due to taxonomic reasons such as an amended circumscription. Thus, the Code also stipulates that when a conserved name competes with one or more names based on different types and against which is not explicitly conserved, the earliest of the competing names should be adopted (Art. 14.3). Thus, when Bennet (1971) suggested the merger of *Pongamia* Vent. (1803) and *Derris* Lour.(1790), the latter was chosen as the correct generic name. *Pongamia pinnata* (L.) Pierre, became *Derris indica* (Lam.) Bennet, in spite of the fact that the generic name *Pongamia* Vent. is conserved. Similarly, when *Mahonia* Nutt. (1818) nom. Cons. is united with *Berberis* L. (1753), the combined genus will bear the prior name *Berberis* L. although *Mahonia* is conserved and *Berberis* is not.

Inspite of various protests from the agricultural botanists and horticulturists who are were disgusted over the name changes due to strict application of the principle of priority, the taxonomists for a long period did not agree upon conserving names at the species level. The constant pressure made and also the discovery of *Triticum aestivum*, was not the correct name of common wheat, prompted by the taxonomists to approve, at the Sydney Congress held in 1981, to accept the proposal, in principle, the conservation of the species names – *nomina specifica conservanda* (probably the most important amendment to the Code) and it was accepted with a severe warning that it is not to be abused and with an explicit restriction to names of species of major economic importance.(Art.14.2). As a result, *Triticum aestivum* L. was the first species name to be conserved at the Berlin Congress in 1987 and the another species name *Lycopersicon esculentum* Mill was also conserved at the same time and this was published in the subsequent Code in 1988.

History of the species *Triticum* **is given below to understand the reason for its conservation** Two species of wheat such as *Triticum aestivum* and *T. hybernum* were described and published on the same date by Linnaeus in his Species Plantarum, in 1753. According to the rules of nomenclature, when two species bearing the same date of publication based on the same type are united, the author who first unites the two taxa has free choice of selecting one as the correct binomial for the plant.

Accordingly it was understood for a long time that, the two species of wheat, *T. aestivum* and *T .hybernum* Linn. were first united by Fiori and Paoletti in 1896, who rightly selected *T. aestivum* Linn. as the correct name for wheat. Later in 1980, it was pointed out by Kerguelen that, the first person who united these two species was actually Merat (1821) and he had selected *T. hybernum* Linn. as the correct name for wheat and not *T. aestivum* as was done by Fiori and Paoletti. This observation would have resulted in dropping of *T. aestivum* Linn. in favour of *T. hybernum* Linn. Therefore, a proposal was made by Hanelt and Schultze-Motel (1983) to conserve the name *T. aestivum* Linn., which is being an economically important cereal crop, and their proposal was accepted at the Berlin Congress to conserve it, and also removed any further problem that would be created to the name *Triticum aestivum* Linn. Therefore, *T. aestivum* Linn. forms the first species to be conserved in the *nomina specifica conservanda* list.

Conserved species list The following is the list of species names that have been considered as conserved species names (nomina specifica conservanda) , in which each name is followed by the (=) sign indicating taxonomic synonym or a (= =) sign indicating nomenclatural synonym and then the binomial against which it has been conserved in the 1993 Code, which was published in the year 1994:

- *Triticum aestivum* L., 1753

  (=)

  Triticum hybernum L., 1753
- *Erica carnea* L., 1753

  (=)

  Erica herbacea L., 1753
- *Lycopersicon esculentum* Mill., 1768

  (= =)

  Lycopersicon lycopersicum (L.) H.Karsten, 1882
- *Fraxinus angustifolia* Vahl, 1804

  (=)

  Fraxinus rotundifolia Mill., 1768
- *Bactris gasipaes* Kunth, 1816

  (=)

  Martinezia ciliata Ruiz et Pav., 1768

  Or

  Bactris ciliata (Ruiz et Pav.) Mart.

5. **Exceptions** Certain classes of names are excluded for the purpose of priority.

10

EFFECTIVE AND VALID PUBLICATION

Keywords Printed matter, effective publication, valid publication, periodicals, manuscripts, electronic media, lithography, seed - exchange list, nomen nudum, latin diagnosis, illegitimate names, latin termination, autonyms, cross reference, author citation, important dates, orthographic error, nomenclatural type, synonym, basionym, fossil fungi, homonyms.

Publication is effective when the general botanical public may come to know of the publication of the concerned taxa.

The conditions for effective and valid publication of names have been provided in Articles 29-45 of the Code. The present Code insists the importance of Effective and Valid publication of names, accordingly two basic conditions to be fulfilled before a properly formulated name can have any status in botanical nomenclature.

Firstly, the name must be published in a medium appropriate to the requirements of the Code, and secondly, the name should be published in accordance with the provisions of the Code. If the first condition is fulfilled, the name is said to be "effectively published" and in the latter case the name is "validly published" and these two conditions actually constitute two different aspects of publication.

Effective publication (Arts.29-31) Effective publication deals with standards applicable to the media in which the names are published and not to the name themselves. The conditions and dates for the names of taxa to be effectively published are dealt in the Code and are as follows:

- Publication of a name is effected only by distribution of printed matter.
- Publication of a name is effective only when the printed matter of the taxon is widely distributed through gift, exchange or sale to the general public, particularly to the botanists, or at least to the botanical institutions with libraries accessible to the botanists and also published in the recognised periodicals or journals related to botany.
- The publication is not effected by communication of new names at a public meeting such as during the annual session of the Indian Science Congress, or by placing of names in collections or gardens open to the public, where visitors may see the names or by the issue of microfilm prepared from manuscripts, type-scripts, or other unpublished materials by publication on-line or by distribution of distruable electronic media like Cassets, CD's etc. offered for sale of printed matter that does not exist.
- It is effective by indelible autograph if it was done before 1st January , 1953. Indelible autograph produced at a later date is not effectively published.

For example, *Salvia oxyodon* Webb. & Heldr. was effectively published in July 1850, in a printed autograph catalogue placed on sale.

- Publication of handwritten material, reproduced for indelible autograph, by some mechanical or graphic process such as lithography, offset or metallic etching is effective before 1.1.1953 and the entirely handwritten account of a new taxon with Latin description is treated as unpublished after this date.
- Publication on or after 1st January, 1953 in trademan's catalogues or non-scientific newspapers, and on or after1st January,1973 in seed-exchange lists, is not effectively published.
- The distribution of printed matter accompanying exsiccatae on or after 1st January, 1953 does not constitute effective publication. Whereas, if the printed matter is distributed independently of the exsiccata, it is treated as effectively published.
- Publication of an independent non-serial work stated to be a thesis submitted on or after 1st January, 1953 to a university or other institute of education for the purpose of obtaining a degree is not effectively published unless it includes an explicit statement referring to the requirements of the Code for effective publication or other internal evidence that it is regarded as an effective publication by its author or publisher. The presence of an International Standard Book Number (ISBN) or a statement of the name of the printer, publisher, or distributor in the original printed version is regarded as internal evidence that the work was intended to be effectively published.

For example, the thesis "*Nasa* and the conquest of South America - Systematic Rearrangements in *Loasaceae* Juss." submitted by Weigend (June 1997 to the Ludwig-

Maximilians-Universität München) is not considered effectively published as it does not include the name of any printer or publisher or distributors' name, or any statement, or an ISBN, that the Code recommend for effective publication.

- The Code strongly recommends that authors should avoid publishing new names and descriptions or diagnoses of new taxa (nomenclatural novelties) in a very short printed matter of any kind. Authors must also avoid publishing new names and descriptions or diagnoses in abstracting journals or in popular periodicals.
- Authors and editors must mention nomenclatural novelties in their summary or abstract, or it can be listed in the form of an index in the publication.
- The date of effective publication of the name is the date on which the printed matter become available for publication.

However, when separates from periodicals or other works placed on sale are issued in advance, the date on the separates is accepted as the date of effective publication, unless it becomes an error.

For e.g. the names of the Selaginella species published by Hieronymus (1912) are considered effectively published on 15th October, 1911, as it is mentioned that the volume in which the paper appeared though dated 1912, it states that the separates appeared on that date.

Valid publication (Arts.32-45) Valid publication refer to both distribution of the name and also with preparation of textual matter prior to distribution.

- One of the conditions for valid publication of a name is that, it should be effectively published

(Nicolson, 1980a) on or after the starting-point date of the respective group.

Therefore, in order to have an effective publication of a name of a taxon, certain conditions are to be fulfilled for a valid publication and are given below:

* A name published without description or diagnosis of the taxon or by a reference to an earlier and effectively published description or diagnosis is not valid and such a name has to be rejected as 'nomen nudum' or 'naked name'.

For example. 1. The genus *"Egeria"* published by Néraud (1826), was without a description or a diagnosis or a reference to an earlier one. Hence the name was not validly published and considered as a naked name.

E.g.,2. The generic name *Ibidium* Salis. (1821) was published merely with four species included in it. When these species were published by Salisburry, he failed to provide description or diagnosis for generic name at the time of publication. Hence, the genus *Ibidium* was not validly published, and it became an illegitimate name that has to be rejected as a naked name.

E.g.,3. The name *Loranthus macrosolen* when first published by Steudel (1843) was without description or diagnosis. Hence, the name became an illegitimate and has to be rejected as a naked name. Therefore this name was considered as an illegitimate name till 1847, when A.Richard (1847) validated this name by supplying it with description.

E.g.,4. Similarly, most of the names published in *"Hortus Bengalensis"* (garden plants of Bengal), by Roxburgh (1814) were without description and diagnosis and are not validly published. But later most of these names were validated by different authors by fulfilling the requirements of the Code.

- ★ A diagnosis of a taxon is a statement by which its author distinguishes the taxon from others.
- ★ The names published on or after 1st January, 1953 must accompany with its full and direct reference to a previously and effectively published description or diagnosis for valid publication.
- ★ At the same time, one should understand that, by giving reference to a description or diagnosis to a name alone will not validate publications published before 1753. A newly described taxon should have points of differences from its allied species.
- ★ A name published with an incorrect Latin termination but otherwise in accordance with the Code is regarded as valid, and the names are changed without the change of author's name and date of publication.
- ★ Autonyms are treated as validly published names from the date of publication (Art.32.7).
- ★ Names in specified rank included in the publications listed as a suppressed works (*opera utique oppressa.* App.V) is not validly published.
- ★ The authors should avoid the adoption of a name which was previously but not validly published for a different taxon.
- ★ In describing or diagnosing the new taxa, the authors should provide illustrations to support proper identification and explanation and also the scale of figures and indication of specimen(s) on which they are based.
- ★ The description or diagnoses of parasitic plants should always be accompanied by the indication of host, especially those of parasitic fungi with designation of scientific names and names should not be in modern languages.

- ★ A combination (except Autonyms) is not validly published unless the final epithet is definitely associated with the name of the genus or species or with its abbreviation by the author.

- ★ A new combination or an avowed substitute published on or after1st Jan,1953 for a previously and validly published name should clearly indicate the basionym (epithet bringing synonym) or the replaced synonym is clearly indicated with a full and direct reference to its author and original publication with page reference and date, on which the new combination is based.

For examples

1. *Selaginella pallescence* (C.Presl) Spring.

   Lycopodium pallescence C.Presl. –Basionym.

2. *Arabidopsis mollissima* var. *glaberrima* (Hk.f.&Th.) Schulz, (1924), is a validly published name, when its basionym, *Sisymbrium molissimum* var. *glaberrima* Hk.f & Th. with full citation is mentioned in the text.

3. The species *Ectocarpus mucronatus* D. A. Saunders when transferred to *Giffordia*, the authors Kjeldsen & Phinney (27 Apr. 1973) cited the basionym with its author but failed to cite the reference to its place of valid publication. Later they (2 Jul. 1973) validated the name *G. mucronata* (D. A. Saunders) Kjeldsen & Phinney by giving a full and direct reference to its place of valid publication of the basionym.

   - ★ A name given to a taxon whose rank is at the same time, contrary to the Code (Art. 5), denoted by a misplaced term is not validly published and such misplaced terms include varieties, species containing genera, and genera containing families or tribes.

For example, 1. "*Section Orontiaceae*" published by Brown (1810) was not validly published , since he misapplied the term "section", to taxa of a rank higher than genus.

- ★ According to the Code, a name is not validly published when it is not accepted by the author in the original publication, or proposed by the author for future acceptance, or cited as a synonym or mentioned as the subordinate taxa, and it also does not apply to names published with a question mark or any other indication of taxonomic doubt published and accepted by the author.

E.g., 1. The species "*Erythrina micropteryx* Poepp." was not validly published because; it was cited as a synonym of *Micropteryx poeppigiana* Walp. (1850), when this species is later on placed under the genus *Erythrina* L., Therefore it is rightly called *E. poeppigiana* (Walp.) O. F. Cook (1901).

- ★ The Code states that on or after 1.1.1953 none of the name is validly published, when two or more different names are proposed simultaneously for a single taxon by the same author, however, this rule does not apply in cases where same combination is used at different ranks.

For e.g.,1. The species *Euphorbia jaroslavii* Polj (1953) was published with an alternative name *Tithymalus jaroslavii* and both are not validly published, however, one of the species *E.yaroslavii* Polj. was validly published by Poljakov (1961) with a change of initial letter of a specific epithet and rejected the other name.

- ★ The new name or combination that have been published on or after1st Jan. 1953 must clearly indicate the rank of the taxon concerned, for valid publication, otherwise the name is considered as not validly published. It should be clearly indicated whether it is a new genus (gen.nov.), new species (sp.nov.) or a new combination (comb. nov.), but if a new name or a combination published before 1st January 1953 without a clear indication of rank is accepted as validly published provided all the requirements for valid publication are fulfilled. However, it can be applied for priority except for homonymy.

For e.g.,1. The groups *"Soldanellae", "Sepincoli", "Occidentales"*, etc., were published without any indication of rank in the genus *Convolvulus* L. by House (1908). The names *C.* [unranked] *Soldanellae,* etc., are validly published but they are not in any definite rank, therefore they have no status in questions of priority except that they may act as homonyms.

For example, 2. In the genus *Carex* the epithet Scirpinae was published as an infrageneric taxon but without specification of any rank by Tuckerman (1843).(C.[unranked] Scirpinae Tuck.). Later it was assigned to the rank of 'section' by Kukenthal (1909). Hence it is cited as Carex sect.Scirpinae (Tukr.) Kuk.

- The names of new taxa of plants, with the exception of algae and all fossils published on or after 1st January 1935 must be accompanied by a Latin description or diagnosis, or by a reference to a previously and effectively published Latin description or diagnosis of the taxon for valid publication. This would make the description of the taxon readily comparable.

For example,1. *Phalaris arundinacea* Linn. Sp.Pl. 55, 1753 has a Latin diagnosis and therefore it is a validly published name.

E.g.,2.*"Schiedea gregoriana"* Degener (1936), was not validly published because, it was not accompanied by a Latin description or diagnosis at the time of its publication. Hence it became an illegitimate name.

- ★ The name of a new taxon of non-fossil algae published on or after 1st January, 1958 is valid only when it is accompanied by a Latin description or diagnosis or by a reference to a previously and effectively published Latin description or diagnosis of it.

For e.g., The alga *Neoptilota* Kylin (1956), when published, was accompanied only by a German description and not Latin at the time of publication. Inspite of it, the name is considered as validly published as it was published before 1958.

- ★ A name of a new taxon of fossil plants published on or after 1.1.1996 is accompanied by a Latin or English description or diagnosis or by reference to a previously and effectively published Latin or English description or diagnosis is treated valid.
- ★ The name of a new species or infraspecific taxon, is acceptable with a citation of a single element to indicate the holotype, and in addition the name of the place where the type specimen is permanently preserved and conserved should be indicated, (Art.37) with citation of author(s), collection number and date, and reference to other details of the type specimen or illustration is required as per the recommendation of the Code.
- ★ The name of a new taxon of the rank of genus or below published on or after 1st January, 1990, is valid only when a type specimen is indicated with one of the words "typus" or "holotypus", or its abbreviation, or its equivalent in a modern language.
- ★ The name of a new species or infraspecific taxon published on or after 1st January, 1990 is valid only when the type specimen or unpublished illustration,

the herbarium or institution in which the type is conserved must be specified.

- The name of a new taxon of a fossil plant of specific or lower rank published on or after 1st January, 1912 must accompany with an illustration or figure with essential characters, in addition to the description or diagnosis or by a reference to an earlier and effectively published illustration or figure is validly published (Art.38).
- The names of hybrids of specific or lower rank with Latin epithet are valid only when they agree with the rules of the names of non-hybrids of same rank (Art.40).

For example, the name "Nepeta × faassenii" Bergmans (1939) was provided with Dutch description and later in 1949 with English description. Hence this name is not valid due to the lack of Latin description. However the name Nepeta × faassenii Bergmans ex Stearn (1950) was validated by supplying it with Latin diagnosis.

- The name of family, subfamily, genus or subdivision of a genus is validly published when accompanied by a description or diagnosis of the taxon or a cross reference (Art.41).
- The name of a genus and a species together is validly published with one description or diagnosis only, provided the genus is monotypic, with no other previous name published on same type. Reference to an earlier description or diagnosis is not acceptable in this case. However, before 1st January, 1908, an illustration with analysis or for non-vascular plants a single figure for identification is acceptable instead of a written description or diagnosis.(Art.42).
- The name of a taxon below the rank of a genus is not validly published unless the name of the genus or

species to which it is assigned is validly published at the same time or earlier. The same rule also holds to specific and other epithets.

For e.g., The genus *Suaeda* was published as a new genus with six species by Forsskal in 1775 with the description and a diagnosis pertaining to the species. But he failed to provide description or diagnosis for the genus. Therefore the name of the genus along with all the species were considered as not validly published.

- The name of a species or infraspecific taxon published before 1st January, 1908 is validly published (Art.44), if it is accompanied only by an illustration with analysis with essential characters.

For example *Panax nossibiensis* Drake (1896) was validly published on a plate with analysis.

- The date of a name is the date of its valid publication and a name published on or after 1st January, 1973 for which the various conditions for valid publication are not fulfilled simultaneously is not validly published unless a full and direct reference is given and a correction in the original spelling of a name does not affect its date of valid publication (Art.45).

For e.g., The correction of the orthographical error in *Gluta "benghas"* L. (1771) to *Gluta renghas* L. does not affect the date of publication of the name even though the correction date was effective only from 1883.

- After 1st January, 2000, when one or more of the other conditions for valid publication have not been met prior to registration, the name must be resubmitted for registration after these conditions have been fulfilled.
- According to St. Louis Code (2000), in the International Code of Botanical Nomenclature

(Art.31.1-2) and also the rule already included in the Tokyo Code, new names of plants and fungi have to be registered in order to be validly published after 1.1.2000.

- Registration on trial basis has already started, on a non-mandatory condition, for a two – year period starting from 1st January, 1998. The Secretariat of IAPT forms the Co-ordinating Centre, which is currently located at the Botanical Garden and Botanical Museum Berlin – Dahlem, Germany. In addition the International Mycological Institute in Egham, UK, also acts as an Associate Registration Centre for the whole of fungi including fossil fungi.
- For purpose of priority only legitimate names are taken into consideration. However, validly published earlier homonyms, whether legitimate or not, shall cause rejection of their later homonyms, unless the latter are conserved or sanctioned.
- All validly published names may not necessarily be legitimate names, similarly all legitimate names may be correct or incorrect according to the different concepts of taxa. As these terms have different connotations in botanical nomenclature, these names should be used very carefully.
- Legitimate names are those that are in accordance with the rules of the Code, whereas others are illegitimate. The correct name of a taxon is the legitimate name that must be adopted for it under the rules (Art.6).

For example 1. Both the generic names *Justicia* Linn. (1737) and *Rostellularia* Reich. (1837) are legitimate, as they were published in accordance with the rules of the Code.

However, selecting the correct name from these two genera depends on the perceptions of the group (taxonomists) who works upon it.For those who accept the genus in the broad sense, inclusive of *Rostellularia*, the correct name is *Justicia* Linn.(1737),as it was published earlier and the other name *Rostellularia* is designated as a synonym.

Similarly, the generic names *Hibiscus* Linn. (Sp Pl. 693. 1753; Gen.Pl.ed. 5. 310:1754) and *Abelmoschus* Medicus (Malv. 46. 1787) are legitimate names as they were published according to the rules. Both the generic names are correct, when the genera are treated separately. In 1824, Augustin de Candolle united both these genera to a single genus *Hibiscus* and this was accepted as followed in most of the earlier Indian floras. Therefore the name *Hibiscus* L. was considered as the correct name with this particular circumscription. However, K.Schumann (1895) and also recent monographers re-established the genus *Abelmoschus* Medicus. The legitimate generic name *Abelmoschus* Medicus may therefore be correct or incorrect according to different concepts of the taxa.

★ The name of a taxon is not validly published, if it is cited merely as a synonym.

★ The description and diagnosis of new taxa, published before 1st January , 1935 are treated as valid, even if they were in any modern language including Japanese, Russian or any other where Roman alphabets are not used.

11

CITATION

Keywords Definition, importance, recommendations, abbreviated names, citation of authors, use of words, examples, later homonym, synonym, infraspecific taxon.

In the First International Botanical Congress held in Paris (1867), a provision was made for citation of authors among other De Candollean rules. The Article 46 of the Code deals with the condition to be followed for citation of authors. It states that "For indication of the name of a taxon to be complete and accurate, it is necessary to cite the name of the authors who first published the name in question". The indication of name of person(s) after the binomial at the time of description and publication is called citation.

Importance of author's citation Though name of authors to any plant does not form the part of the name of the plant, as such, but the citation has the following purpose:

1. The citation of author expedites locating the original plant description, which helps to determine the type and date of publication of the taxon.

2. It helps to verify the order and accuracy of the date of publication.
3. In case of any doubt in understanding of any character of the plant, the author citation is very helpful in checking the original description given by the original author of the plant.
4. It can aid the botanists in tracing the transference of species from one genus to another.
5. Author citation is provided for purpose of brevity, precision and for identification of the plant from the original author.
6. It allows the distinction between the two names and also to find out the legitimate name.
7. It also gives us some idea about the place from where the plant specimen was collected and identified.
8. It helps to identify the correct name and also error made by author can be detected in the same genus.
9. Author citations can indicate the Bibliographic references, which help in the identification of homonyms. For e.g. *Utricularia caerulea* Linn. and

   Utricularia caerulea Clarice are two names referring to two different taxa, but it would not have been possible for one to recognise this as a homonym without the citation of author's names.

Recommendation for author citation

1. The name of the author(s) who first published must be cited after the name of the plant.

   E.g. *Cocos nucifera* Linn.

2. It is recommended that the name of the authors are cited as full name or abbreviated if it is too long.

   E.g. 1. *Adhatoda vasica* Nees

   E.g. 2. *Bambusa arundinacea* Willd.

Abbreviated names of different authors and their full name

Abbreviated name	Full name
Linn.\ L.	Carolus Linnaeus
Ait.	Aiton
All	Allioni
Roxb	Roxburgh
Buch.- Ham	Buchanam Hamilton
Wall	Wallich
Bl	Blume
R.Br	Robert Brown
Benth	G. Bentham
Lamk	JP Lamark
Carr	Carriere
Pers	C.H.Persoon
Willd	Carl Willdenow
DC	A.P. de Candolle
HBK	Humboldt, Bonpland and Kunth
Koch	Karl Koch
Mill	Miller

When two or more than two persons of the same family members are involved in the publication of names, which is necessary to distinguish the relatives, the following methods are adopted:

De Candolle family Father Augustin de Candolle is cited as DC., and his son Alphonse as A.DC., and their grandson Casimir as C.DC.

Carolus Linnaeus family Father Carolus Linnaeus is abbreviated as Linn. or L., and his son Linnaeus is cited as

Linn.f., where the small letter 'f' represents the Latin word 'filius' meaning the son.(son of Linnaeus.)

William family Father William Hooker name is cited as Hook. and his son Joseph Dalton Hooker as Hook. f.,

3. The name of the authors are never underlined or typed in italics.

   E.g. *Carica papaya* Linn.

4. In all the taxonomic work, it is very important to cite the authors of scientific names.

5. If a taxon is of garden origin, citing the name should be ascribed to Hort. (hortulanorum) and connected to the name of the concerned author who published it by adding a word "ex" as cited below:

   E.g. *Gesneria donklarii* Hort. ex Hook.

Publication of names The plant name may be published either by a single person or by two or more than two persons. Accordingly the citation of author(s) is written as given below:

★ **Publication of name by an author** When the plant name is first identified, described, named and published by an author, the citation is

  Portulaca oleracea L., Sp.Pl. 445. 1753.

In the above citation, the plant *Portulaca oleracea* was first identified, named, described and published by the author Carolus Linnaeus in his work *Species Plantarum* on page 445, in the year 1753.

★ **Publication of name by many authors** The plant name may be jointly published by two or more than two authors for various reasons. Accordingly different words such as et or &, et al, in, ex, emend, square brackets, nom.cons., nom.nud., auct.non., pro.part., sensu amplo., pro.syn., parenthesis etc. are used in citations to indicate the different purpose, so as to make the reader understand, what kind of changes are made by different authors in the taxonomic work of different plant species.

The examples of different author citations are as follows:

a. Use of et or &:

When a name of a taxon is published jointly by two authors, both the names are connected by the word 'et' or an ampersand (&), as cited below:

E.g. *Pennisetum typhoides* Stapf et Hubbard.

or

Pennisetum typhoides Stapf & Hubbard.

In the above citation the word 'and' should not be used between the authors' names since the name of the plant is in Latin.

b. Use of *et al*:

When a name is published jointly by more than two authors, the citation is restricted to that of the first senior author followed by the word *et al*., (*et al,* means, and others).

E.g. *Indotristicha tirunelveliana* Sharma *et al*.,

The above plant name was published by Sharma, Karthikeyan and Shetty (Bull.Bot. Surv.of India. 16:157-161. 1974). Hence the name of the senior author Sharma is followed by the word *et al*., in the citation.

c. Use of in:

When a name of a taxon with a description or diagnosis supplied by one author is published in another author's work, the word 'in' should be used to link the names of these two authors.

E.g. *Osbeckia wynaadensis* C.B.Clarke in Hook.f., Fl. Brit. India 2:521,1879.

In the above citation, the author C.B.Clarke actually published his new species in the Flora of British India, the work of Sir J. D. Hooker.

d. Use of ex:

When an author who first validly publishes a name ascribes it to another author who actually described and proposed the name of the plant, but failed to publish it, the citation is written as the binomial of the plant,which is followed by the validly published author's name alone or the binomial is followed by the author who first proposed the name of the plant, followed by the word 'ex' and then the name of the person who actually published it.

E.g. *Cassia montana* Roth.

or

Cassia montana Heyne ex Roth.

In the above citation, author Heyne first proposed the name *Cassia montana* but he did not publish it. But later the plant was validated by the author Roth by validly publishing it. The meaning of 'ex' is "validly published by".

Similarly, *Hopea wightiana* Wight & Arn.

In which the plant was validly published by Wight & Arn. Therefore in the citation, the binomial is followed by the persons who validly published it, i.e. Wight & Arn.is given in the citation or the citation is also indicated with the use of the word 'ex' as *Hopea wightiana* Wall ex Wight & Arn. In which Wallich who first proposed the name *Hopea wightiana* in his catalogue (1828- 1849) by simply listing it without giving any description. Later Wight & Arnolt provided the description for the same name and validated the name by validly publishing it.

e. Use of emend:

When the original description of a taxon is incomplete and subsequently corrected the diagnosis or the circumscription of a taxon is altered retaining the same type, the citation of the original author is retained but followed by the word 'emend' and then the name of the author who did the correction (Art.47). This is an important element of revision of any taxonomic work.

E.g. *Phyllanthus* Linn. emend Mull - Arg.

The word 'emend' comes from the word 'emendavit' – means "person making correction".

f. Use of square brackets:

The Montreal Code of 1959 insists that "the citation of an author who published the name before the introduction of the binomial system of nomenclature 1753 is indicated preferably between square brackets.

When a name of a taxon is published before the starting point of the group is concerned, square brackets are used in the citation. For e.g. the generic name *Lupinus* was defectively published by Tournefort in 1719, but this date was found to be earlier than the date of botanical nomenclature based on *Species Plantarum* 1753 of Linnaeus, hence the correct citation for the genus is *Lupinus* [Tourne.] Linn.

g. Use of *nom. cons*. (nomina conservanda):

If a generic or specific name is accepted by the committee for conservation, the abbreviated words *nom. cons.* should be added to the citation as given below:

E.g. *Tectona* L.f. Suppl. 20, 151. 1781. *nom.cons.*

h. Use of 'nom. nud'. (nomina nudum):

A name published without description is called a naked name. When such name is cited, its status should be indicated by adding the words 'nom. nud'. after the authors name.

E.g. *Convolvulus adpressus* Wall. Cat. No. 1424, 1828, 'nom.nud'.

The author Wallich in his catalogue (1828-1849) published many new plant names without any description. All these names are not valid, therefore cited with the words 'nom. nud'. According to the Code, naked names are not valid and are to be rejected.But most of these names were later on validated by giving description and publication by different authors. Use of 'auct. non'. :

In the citation, a misapplied name is indicated by the words 'auct. non'. after the binomial, it is then followed by the name of the original author and the bibliographical reference of the misidentified name is also given.

E.g. *Phyllanthus fraternus* Webster, Contr. Gray Herb. 176: 53. 1955. *Phyllanthus niruri* auct. non. L. 1753. Hook.f. Fl. Brit. India 5:298.1887.

In most of our Indian Floras, this plant has been treated as *Phyllanthus niruri* L., which is locally called kezhanelli in Tamil language, and given for the treatment of jaundice. But Webster's work proved that, the plant distributed in India is actually *Phyllanthus fraternus* but not *Phyllanthus niruri* as described by Linnaeus. Hence, it is an example of misapplied name and the words 'auct.non'. L. in the citation mean that the plant treated as *Phyllanthus niruri* L. by many authors, is not the species *P. niruri* which was described by Linnaeus, but it is *Phyllanthus fraternus* Webster. Hence, a misapplied name must be indicated by the words 'auct. non'. as cited above.

i. Use of 'pro.parte' (pro. parte means partly or in part, (p.p)): When an alteration of the circumscription of a taxon is divided into two or more taxa, in the citation of the newly formed taxa, the misapplied original name is indicated by adding the words 'pro parte' or p.p.

For e.g. 1. *Maba neilgerrensis* Wight, 1850.

Maba buxifolia sensu Clarke in Hook. f. Fl. Brit. India 3: 551. 1882. p.p. non Pers. 1807.

In the above citation it is very clear that the words p.p. indicates that a part of the description of *M. buxifolia* which was given by Clarke in Hooker's Flora of British India, actually pertains to the other species of the same genus *Maba neilgerrensis* Wight.

Therefore, *M. buxifolia* sensu Clarke means *Maba buxifolia* as considered by Clarke; and non (not as treated by) Pers. 1807 means *M . buxifolia* as treated by Clarke is not that of the true species of *M . buxifolia* Pers. 1807.

j. Use of sensu amplo (s. ampl.) and sensu stricto (s. stri.):

(The sensu amplo or sensu lato means in broad or wider sense and sensu stricto means in the narrow sense).

For e.g. the genus *Eugenia* L. in the broad sense (sensu amplo) includes species of *Syzygium, Cleistocalyx, Eugenia* – proper etc. That means the circumscription of the genus (*Eugenia*) is widened to accommodate the other allied genera. Therefore, when *Syzygium* and *Cleistocalyx* are treated as distinct genera from *Eugenia* – proper, the latter is cited as *Eugenia* L. sensu stricto.

k. Use of 'pro. syn'.:

It is also recommended in the Code that, when an author has published as a synonym, a manuscript name of another author, the word syn. should be added.

For e.g., *Descuriania sophia* (L.) Webb. Benth. Pflanz enf., 3 (2): 192.1891. syn. *Sisymbrium sophia* L.,Sp.Pl.659.1753.

The words 'pro.syn'. refers to synonym. When the name is published as a synonym in the citation, the words 'pro.syn' should be used.

For e.g. when an author has published as a synonym, a manuscript name of another author, the word 'ex' is added in the citation to link the name of the two authors.

For e.g., the species *Hedyotis evalata*, a manuscript name of Beddome published by Gamble as a synonym of *Oldenlandia evalata* Gamble, should be thus cited as:

Hedyotis evalata Bedd. ex Gamble in Kew Bull. 1919: 405. 1919, pro.syn. However it will not be considered as a valid publication.

l. Use of parenthesis:

Article 49 of the Code states that, "when a genus or a taxon of a lower rank is altered in rank or position but retains its name or epithet, unless it becomes a tautonym or preoccupied in the same genus, the author of the basionym who first published the name as a legitimate name or epithet bringing synonym must be cited in parenthesis and the name of the author(s) who effected the alteration is cited outside the parenthesis". The same rule is also applicable when a taxon of the lower rank than genus is transferred to another genus or species with or without alteration of rank.

The same rule also applies when a subdivision of a genus (Art.54) or a species (Art.55) is transferred to another genus or placed under another generic name.

The same rule also applies when infraspecific transfer or change of rank is effected. The citation has resulted due to these changes is an example of double citation.

A. Change in taxonomic position:

1. *Lycopodium pallescens* C.Presl. when transferred to the genus *Selaginella* by Spring, it is cited as,

Selaginella pallescens (C.Presl) Spring.

In the above example, after the binomial, the name of the author (C.Presl.) who first validly published the name must be cited in parenthesis, and the name of the author who effected the transfer (Spring) is cited outside the parenthesis. In the above citation, *Lycopodium pallescens* presl is the epithet bringing name or basionym.

2. *Cyanodon dactylon* (Linn.) Pers. is based on the basionym *Panicum dactylon* Linn. The original name of the species has been described and published by Linnaeus.

3. *Amischophacelus axillaris* (L.) R. Rao & Kam. 1966. is based on the basionym *Commelina axillaris* L. 1753.

4. *Mesia serrata* Gaertn, was transferred to the genus *Gomphia* by Kanis, therefore it is written as *Gomphia serrata* (Gaertn.) Kanis.

5. *Althea rosea* (L.) Cav., originally called *Malva rosea* L.which was validly published by Linnaeus, and later transferred to the genus *Althea* by Cavanilles. Hence, it is cited as given above.

6. When *Oxalis sensitivum* L. was transferred to the genus *Biophytum* by A.P. de Candolle, it becomes *Biophytum sensitivum* (L.) DC.

7. *Saponaria* sect. *Vaccaria* DC. when transferred to *Gypsophila*, becomes *Gypsophila* sect. *Vaccaria* (DC.) Godr.

8. *Limonia aurantifolia* Christm. when transferred to the genus *Citrus* becomes *Citrus aurantifolia* (Christm.) Swingle.

9. *Hedysarum velutinum* Willd. when transferred to the genus *Desmodium* becomes *Desmodium velutinum* (Willd.) DC.

B. Change in taxonomic rank:

i. The species *Sida retusa* L. when reduced to a subspecies of *Sida rhombifolia* L. by Boiss, it is cited as *Sida rhombifolia* subsp. *retusa* (L.) Boiss.

ii. *Draba rupestris* var. *altaica* C.A. Mey, when raised to the rank of a species by Bunge, is written as *Draba altaica* (C.A.Meyer) Bunge.

iii. *Medicago polymorpha* L. var. *orbicularis* was first published by Linnaeus, which was later elevated to the rank of a species by Allioni and thus it becomes *Medicago orbicularis* (L.) All.

iv. *Citrus aurantium* var. *grandis* L., when raised to the rank of a species becomes *Citrus grandis* (L.) Osbeck. In the above citation, L. in the parentheses is the person who first indicated *grandis* as var. under the species *Citrus aurantium*, while the author Osbeck who raised the variety *grandis* of Linnaeus into species under the genus Citrus. Hence the citation appears as it is written above.

v. *Alysicarpus nummularifolius* DC. when reduced to variety, it becomes *Alysicarpus vaginalis* var. *nummularifolius* (DC.) Baker.

C. Change in taxonomic rank and position:

a. *Acacia nilotica* ssp. *indica* (Benth.) Brenam. 1957.

The basionym is *Acacia arabica* var. *indica* Benth. 1842.

In the above example, the double citation is due to the change in taxonomic rank as well as in taxonomic position. i.e. the variety *indica* of Benth has been raised to the rank of a subspecies (change in rank) and also the position has also been changed from the species *Acacia arabica* Willd. to *Acacia nilotica* (L.) Willd. ex Dalile (change in taxonomic position).

b. *Fumaria bulbosa* var. *solida* L. when it was raised to the rank of specific level by Miller, it becomes *Fumaria solida* (L.) Miller. But the name of this species when transferred to *Corydalis* Clairv, it is cited as *Corydalis solida* (L.) Clairv, but not *Corydalis solida* (Miller) Clairv. This is because, the name *solida* was named by Linnaeus, not Miller.

D. Change of binary names:

According to the Stockholm Code 'if the status of a taxon bearing a binary name is changed from that of a species to that of a hybrid category of corresponding rank or vice versa, the name of the original author is cited and may be followed by an indication of the changed rank in the parentheses (Art.50)".

For e.g. 1. *Stachys ambigua* Smith which was published as a species, if considered as a hybrid, it is cited as *Stachys × ambigua* Smith (pro sp.).

2. Similarly, *Salix × glaucops* Anderson was published as the name of a hybrid, when changed to species level, the name is cited as *Salix glaucops* Anderson (pro hybr.).

E. Naming of an infraspecific taxon:

When naming an infraspecific taxon, the authority is cited for both specific epithet and infraspecific epithet.

For example *Acacia nilotica* (Linn.) Del. ssp. *indica* (Benth.) Brenam. But its autonym for the infraspecific epithet, does not have the author's name, as it is based on the same type as the species. Hence it is cited as e.g. *Acacia nilotica* (Linn.) Del. ssp. *nilotica*.

F. Publication of later homonym:

When an author adopts an existing name but excludes its original types, it is regarded that he/she has published a later homonym and should be ascribed to him/her (Art.48).

For example *Amorphophallus campanulatus* Decaisne, was apparently based on *Arum campanulatum* Roxb. However, the type of the latter was excluded by Decaisne, and the name is to be cited as *Amorphophallus campanulatum* Decaisne, and not as *Amorphophallus campanulatum* (Roxb.) Decaisne.

12

RETENTION OF NAMES AND EPITHETS OF TAXA

Keywords Taxa, revision, retention of names, division, transfer, examples.

In order to avoid confusion and also to overcome difficulties in retaining, choosing and rejection of certain names, the taxonomic studies need realignment of taxa, some of them have to be divided, still others to be united and others have to be remodelled. Accordingly the rules are framed and adopted from time to time and published by the International Botanical Congress.

The rules for the retention of names or epithets of taxa are dealt in Arts. 51-56.

An alteration in the diagnostic characters or in descriptions or in the circumscription of a taxon by new author(s) does not warrant a name change.

1. Retention of name in revisionof taxa

For example 1. The genus *Myosotis* L. was revised by R . Brown, who made some changes in the diagnostic characters of the original genus of Linnaeus, but the generic name *Myosotis* must be retained, since the type of *Myosotis* L. remains in the same genus. Hence the citation is written as given below using the word 'emend' before the name of the person who did the alteration.

Myosotis L. emend. R . Brown

E.g. 2. The genus *Gomphandra* Wall. ex Lindl. was revised by Beccari, who did some alterations in the diagnostic features and circumscription of the genus, since the type of *Gomphandra* remains in this genus, the generic name must be retained. Therefore, it is cited using the word 'emend' as given below:

Gomphandra Wall. ex Lindl. emend Becc.

2. Retention of name of taxa when divided

When a genus is divided into two or more genera, the original name must be retained for one of them which include the type species (Art.52). The same rule is also applied to species and infraspecific taxa (Art.52,53).

For example 1. The genus *Dicera* J. R & G. Forster (1776) was divided by Rafinesque (1838) into two genera such as *Misipus* and *Skidanthera*, which was contrary to rules and consequently the original name *Dicera* was retained as one of the genera for the part which contains the lectotype *Dicera dentata*.

E.g. 2. When the genus *Ludwigia* was divided into two genera, the part which included the type species retained the original generic name *Ludwigia* and for the other one, the name *Fissendocarpa* (Haines) Bennet was given.

Similarly when a species is divided into two or more species, the original specific epithet must be retained for one of the species, which includes its type specimen.

For example1. The species *Heliotropium indicum* Linn. is divided into two species. The one which includes the type retains the original name *Heliotropium indicum*

Linn. and the other is given with a new name (specific epithet), *Heliotropium keralense* Manilal et Sivarajan.

E.g.2. *Lychnis dioica* L. was divided by Miller into two species, which were named as *Lychnis dioica* L. emend Mill. and *Lychnis alba* Mill.

E.g. 3.The species *Potulaca pilosa* when divided into two subspecies, the taxon which includes the type of the species receives the name *Portulaca pilosa* subsp. *pilosa*.(Geesink,1969), while the other is known as subsp. *grandiflora* and is cited as *P. pilosa* ssp. *grandiflora*. The former subspecies consists of two varieties, the one which contains the type material of the species is *Portulaca pilosa* ssp.*pilosa* var.*pilosa* and the other is named as var. *tuberosa* (Sivarajan,1981a) and cited as *Portulaca pilosa* ssp.*pilosa* var.*tuberosa*

3. Retention of name of taxa on transfer

When a subdivision of a genus is transferred to another genus without change of rank, its original name if legitimate must be retained.

Similarly, when a species is transferred to another genus or placed under another generic name of the same genus without a change of rank, the specific epithet if legitimate, must be retained, unless one of the following barriers exist (Art.55 &56):

i. an earlier legitimate specific epithet is available (Art.55).

ii. the resulting binary combination name is a later homonym or a tautonym.

For example, 1. The plant *Saponaria* sect. *vaccaria* DC. when transferred to *Gypsophila*, it becomes *Gypsophila* sect. *vaccaria* (DC.) Godr.

E.g. 2. *Hydrocotyle asiatica* L. when transferred to the genus *Centella* must be called *Centella asiatica* (L.) Urban.

E.g.3. *Ecbollium linneanum* var. *laetevirens* Clarke is transferred to the species *E.viride*, becomes *E.viride* var. *laetevirens* (Clarke) Raizada.

E.g. 4. *Melochia cordata* Burm.f. is transferred to the genus *Sida* becomes *Sida cordata* (Burm.f.) Borssum waalkes.

E.g.5. When *Antirrhinum spurium* L. was transferred to the genus *Linaria* by Miller, it becomes *Linaria spuria* (L.) Mill.

E.g.6. *Spergula stricta* Sw. (1799) when transferred to *Arenaria,* is rightly called *Arenaria uliginosa*. Schleicher ex Schlechtendal (1808), because *Arenaria stricta* Michx (1803) is already existing in the literature. But when transferred to the genus *Minuartia* the epithet *stricta* is retained as *Minuartia stricta* (SW.) Hiern (1809).

E.g.7. *Helianthemum italicum* var. *micranthum* Gren. and Gordon when transferred to *Helianthemum penicillatum* Thibaud ex Dunal as a variety, it must retain its varietal epithet, thus it is cited as *Helianthemum penicillatum* var. *micranthum* (Gren & Gordon) Grosser.

13

CHOICE OF NAMES

Keywords Articles, number, selection of names, taxa, dates, union, non- fossil, fossil taxa, fungi, different ranks, higher rank, pleomorphic cycle, change of names.

Articles 57-58, of the Code deal with the stipulations to be adopted for choosing the correct name, when the taxa of the same rank are united, and when the rank of a taxon is changed is given in Arts.60-61.

1. **Choice of name when taxa of the same rank with different dates united** When two or more taxa of the same rank with different dates are united, the oldest legitimate name should be adopted, unless it contravenes other provisions of the Code (Art.57).

For e.g. i. K.Schumann (Engler & Prantl, Pflanzenfam. III,6:5. 1895) combined three genera such as, *Sloanea* L.(1753), *Echinocarpus* Blume (1825) and *Phoenicosperma* Miq. (1865) in a single genus, and he rightly adopted the oldest of these three generic names, *Sloanea* L. for the resulting genus.

E.g. ii. When *Sida rhombifolia* Linn. (1753) is united with *S . rhomboidea* Roxb. ex Flem. (1810), the correct name is *Sida rhombifolia* Linn. being the earliest.

E.g. iii. When *Derris* Lour. (1790) and *Pongamia* (1803) are merged, the correct name for the genus is *Derris* Lour. as it is the oldest legitimate name.

2. **Choice of name when taxa of the same rank with same dates united** According to the Code, the author who first unites taxa bearing names for epithets of the same date must choose any one of them as a correct name and his choice should be strictly followed.

For e.g. 1. Linnaeus published *Dentaria* L. (1753) and *Cardamine* L. (1753) on the same date that is 1753. Crantz (1769) when united these two genera, selected the name *Cardamine* L. for the combined genera and this name is to be retained and his choice should be strictly followed.

E.g. 2. Robert Brown for the first time when united the two species of Linnaeus, *Waltheria* such as, *Waltheria indica* L. (Sp.Pl.673, 1753) and *Waltheria americana* L. (Sp. Pl.673, 1753), which was published on the same date-1st May in Species Plantarum 673. 1753, he adopted the name *W.indica* L. for the combined species and his selection should be retained.

E.g. 3. The generic names *Thea* L.(1753) and *Camellia* L. (1753) are treated as having been published simultaneously in May 1753 by Linnaeus in his Species Plantarum. Sweet (1818) who was the first person to unite the two genera chose *Camellia* for the combined genus, citing *Thea* as a synonym.

However, in the Code it is recommended that the name adopted should be the one which was accompanied by the description of the species and ii. had many species.

3. **When taxa of non-fossil and fossil plants of the same rank united** When a taxon of non- fossil plants and a fossil taxon of the same rank except that of algae are united, priority is given to the correct name of the non-fossil (Art.58).

For e.g. when *Platycarya* Sieb & Zucc (1843), a non-fossil genus and *Petrophiloides* Bowerbank (1840), a fossil genus were united, *Platycarya* was adopted as the name for the combined genus, although *Petrophiloides* was the earlier one.

4. **Name of Fungi with pleomorphic life cycle** According to Art.59, in ascomycetous and basidiomycetous fungi the adoption of the correct name is the earliest legitimate name of a species covering all forms and typified by an element having ascospores, basidiospores, teliospores or other basidium bearing organ (teleomorph) and accompanied by diagnosis or description. However this rule is not applicable to the lichen forming fungi.

When a new form of fungus is described, it is published as a new taxon whose name has a teleomorphic cycle type, or as a new anamorph (anam.nov.) whose name has an anamorphic type.

For e.g. The name *Penicillium brefeldianum* Dodge, based on teleomorphic and anamorphic material, is a valid and legitimate name of a holomorph, inspite of the attribution of the species to a form genus. It is legitimately combined in a holomorphic genus as *Eupenicillium brefeldianum* (Dodge) Stolk & Scott. *Penicillium brefeldianum* is not available for use in a restricted sense for the anamorph alone.

5. **Choice of name when the rank of a taxon is changed** When the rank of a genus or an infraspecific taxon is changed the correct name or epithet is the

earliest legitimate one available in the new rank. A name or an epithet will have no priority outside its own rank.

For example. 1. *Cyperus iria* var. *rectangularis* Kukenth (1935) when raised to the specific rank is correctly called *Cyperus alulatus* Kern (1952) and not *Cyperus rectangularis* (Kukenth.) Bennet (1969). This is because the specific epithet *rectangularis* was already available (preoccupied) in the same genus *Cyperus*.

E.g. 2. *Magnolia virginiana* var. *foetida* L. (1753) when raised to the specific rank is rightly called *Magnolia grandiflora* L.(1759), but not *Magnolia foetida* (L.) Sarg. as the epithet *foetida* was preoccupied in the genus *Magnolia*. Hence a new specific epithet *grandiflora* is adopted.

E.g.3 When a section or a subgenus is raised in rank to a genus or reverse change is effected, the original name or epithet is retained. Similar is the case with infraspecific taxon where original epithet is retained provided it is in accordance with the Code.

According to Sydney Code, the autonym will get the priority over names of the same date that automatically established it (Art.57.3).

For example, when *Heracleum sibiricum* Linn. (1753) included *H. sibiricum* subsp. *lecokii* (Godron & Gren.) Nyman (1879), and also the subsp, *sibiricum* (1879) was automatically established at the same time. When *H. sibiricum* is included under *H . sphondylium* Linn. (1753) as a subspecies, the correct name of the subspecies is *H . sphondylium* subsp. *sibiricum* (Linn) Simonkai (1887), not subsp. *lecokii*.

However, it is recommended that if there is no earlier name or epithet available in the new rank, the original name or epithet may be retained unless it is contrary to the Code.

6. **Choice of name when taxon of higher rank is changed** When a taxon of a rank is higher than the genus and not higher than the family or subdivision of the family is changed and no legitimate name is available in the new rank, the stem of the name is retained and only the termination is altered (-aceae, -oideae, -eae, -inae) unless it is contrary to the Code:

E.g. The subtribe *Drypetinae* Pax (1890) of *Euphorbiaceae* was raised to the rank of tribe by Hurusawa and it became *Drypeteae* (Pax) Hurusawa.

14

REJECTION OF NAMES AND EPITHETS

Keywords Definition, legitimate names, illegitimate names, examples, tautonyms, synonyms, superfluous names, homonyms, naked names, basionyms, ambiguous names, confused names, doubtful names, monstrosity, examples.

Articles 62 - 72 of the Code furnish the conditions under which names and epithets are to be rejected. A legitimate name is one which is published in accordance with the provisions of the Code. While the other names which are contrary to the rules are called illegitimate names, which are to be rejected. At the same time "a legitimate name or epithet must not be rejected merely because it is inappropriate or disagreeable or an another name is preferable or a better known or because it has lost its original meaning" (Art.62).

E.g. 1. The name *Scilla peruviana* L.(1753) must not be rejected simply, because this species is not found in Peru.

E.g. 2. The generic name *Richardia* L.(1753), which is named in honour of a Botanist, Richardson, is not to be changed to *Richardsonia* as was done by Kunth (Mem. Mus. Hist. Nat. 4 : 430, 1818).

The following types of names are to be rejected and are considered as illegitimate.

1. **Tautonyms**: These are the names where the specific epithet exactly repeats the generic name.

   E.g. *Malus malus, Linaria linaria* etc. are to be rejected. However, *Bambusa bambos* and *cajanus cajan* are not tautonyms, as these specific epithets do not exactly repeat the generic names.

2. **Synonyms**: These are different names used for the same taxonomic group or taxon.

   According to the Code, no taxa can have the same name and a plant or group can have only one valid name and all others are considered as synonyms. They are of two types: *Nomenclatural* and *taxonomic synonyms*. Nomenclatural synonyms are names based on the same type. Thus, *Miliusa tomentosa* (Roxb.) Sinclair, *Saccopetalum tomentosum* (Roxb.) Hook.f. & Thomas. and *Uvaria tomentosa* Roxb. are names based on the same type and are regarded as nomenclatural synonyms.

   Taxonomic synonyms are names based on various types considered to belong to the same taxon. Thus, *Hibiscus vitifolius* L. (1953) and *Hibiscus obtusifolius* Willd. (1801) are names based on two different types. A taxonomic judgement led to the proof that both the types are entirely identical. Here, *H. obtusifolius* becomes a taxonomic synonym of *H. vitifolius*.

3. **Superfluous name** (Nomen superfluum, abbreviated as nom. Superfl.): A name is illegitimate and is to be rejected if it was nomenclaturally superfluous when published. i.e. a name given to a taxon for which another name is already available.

i. The species *Chrysophyllum sericeum* Salisb. (1796) is a superfluous name because this taxon already has a valid name *Chrysophyllum cainito* L.

ii. *Picea excelsa* (Lam.) Link is illegitimate, since this name is based on a superfluous name *Pinus excelsa* Lam., 1778 for " Linn. 1753. The legitimate combination under *Picea* is thus *Picea abies* (Linn.) Karst. (1880).

iii. *Elaeodendron* Jacq. (1782) is an illegitimate name, as it was a superfluous name for *Cassine* L. (1753) and the two genera are having the same circumscription. Hence, the name *Elaeodendron* Jacq (1782) published at a later date must be rejected.

4. Homonym: A name must be rejected, if it is a later homonym. i.e. a name if it is spelled exactly like a name previously and validly published for a taxon of the same rank based on a different type.

For e.g. 1. *Astragalus rhizanthus* Boiss (1843) is a later homonym of *Astragalus rhizanthus* Royle (1835) which was published earlier.Therefore A. *rhizanthus* Boiss (1843) must be rejected.

According to the Code each taxon can have only one correct name, and the Code similarly does not allow the same binomial to be used for two different species. If such names are proposed to plants they constitute homonyms. The one published at an earlier date is termed as earlier homonym and that at a later date is referred as the later homonym. The Code therefore rejects the later homonyms even if the earlier homonym is illegitimate.

For e.g. 2. *Zizyphus jujuba* Lamk. 1789 had long been used as the correct name for the cultivated fruit jujube. Later this name was ascertained to be a later homonym of a related species. *Z. jujuba* Mill. 1768. Hence, the binomial *Z. Jujuba* Lamk . 1789 is thus rejected and jujube is correctly named as *Z. mauratiana* Lamk. 1789.

E.g. 3. The earliest name for almond is *Amygdalus communis* Linn. 1753, when transferred to the genus *Prunus,* the name becomes *Prunus communis* (Linn.) Archangeli, 1882 became a later homonym of *P. communis* Huds. 1762, which is a species of plums. Therefore *Prunus communis* (Linn.) Archangeli, 1882 must be rejected

E.g. 4. The name *Banalia* Moq. (1849) given to the genus of the family *Amaranthaceae*, is a later homonym of *Banalia* Rafin. (1840), a name previously and validly published for a genus of the family *Euphorbiaceae*. *Banalia* Moq. (1849) is therefore rejected as a later homonym, and later on it was renamed as *Indobanalia* by Henry & Bela Roy. (Bull. Bot. Surv. India 10: 274, 1968 (1969), as it was based on different type.

5. **Naked names** (Nomen nudum, nom.nud.)**:** The names that are published without any description are called naked names or nomina nuda. For e.g. many names published by Wallich in his Catalogue (1812) (Wall. Cat.) and Roxburgh in his ' Hortus Bengalensis' were naked names without accompanying any description. These names are to be rejected unless and until validated by later authors by providing description.

   E.g.1. *Cerasus cornuta* Wall ex Royle

   2. *Odina wodier* Roxb.

   3. *Erigeron asteroids* Roxb.

   4. *Centranthera humifusa* Wall.

6. **Basionym** When a name, is required by a change in position or rank of a taxon, is taken from a previous name for the taxon, is the basionym of the new name. It is otherwise called epithet- bringing synonym. For example *Rauvolfia serpentina* (L.) Benth. ex Kurz. was

a combination based on the basionym *Ophioxylon serpentinum* L., which is an epithet-bringing synonym.

7. Ambiguous name (nomen ambiguum, Abbre: nom. ambig.): A name is rejected, if it is used in a different sense by different authors, and becomes a source of permanent error.

For e.g. the species of Rose, *Rosa villosa* Linn. is rejected, because it has been applied to several different species and has become a source of error.

8. Confused name (Nomen confusum, nom.confus.) A name is rejected if it is based on a type consisting of two or more entirely discordant elements, which is difficult to select a satisfactory lectotype.

For e.g, the characters of the genus *Actinotinus*, established by Oliver (1888) were based on a specimen derived from two genera such as *Viburnum* and *Aesculus*. But Oliver's type specimen was received from a Chinese collector who actually inserted the inflorescence of a *Viburnum* into the terminal bud of an *Aesculus* . Hence, the name *Actinotinus* which was proposed by Oliver becomes a confused name and must be rejected.

9. Doubtful name (Nomen dubium, nom.dub.) A name is rejected, if it is dubious, i.e. a name of uncertain application, because it is impossible to establish the taxon to which it should be referred. For e.g. Linnaeus (1753) attributed the name *Rhinanthus crista-galli* to a group of several varieties, which he later described under separate names, rejecting the name *R . crista-galli* Linn. Although, this name was used continuously for a period until Schwarz (1939) finally listed this name as nomen dubium, and the name was finally rejected.

10. Name based on Monstrosity A name should be rejected, if it is based on a monstrosity and such name had no status in botanical nomenclature until the publication of the Leningrad Code (Stafleu &Voss,1978). For e.g. the generic name *Uropedium* Lindl., 1846 was based on a monstrosity of the species now referred to as *Phragmidium caudatum* (Lindl.) Royle, 1896. The generic name *Uropedium* Lindl. stands to be rejected, but these names are considered for rehabilitation.

Similarly, the species *Artocarpus ponga* Dennstedt, was based on a monstrosity and considered as an illegitimate name until recently , and thus reinstated with the combination *Hopea ponga* (Dennstedt) Mabberly (1979).

In the same way *Viscum verticillatum* Linn.(type: Sloane Herb. Vol.6.Fol. 110, BM) was also based on a specimen infested by Smut that caused 'Witches broom' of the inflorescence. With the mandate of the current Code, Nicolson & Jarvis (1984) have reinstated the epithet with the correct name of the species as *Cissus verticillata* (Linn.) Nicolson & Jarvis.

Baker (in Hook.f., Fl. Brit. India 2:282.1878) published *Bauhinia divergens* based on Griffith's collection from Burma. Recently, Larsen & Larsen (1979) have studied the type collection and have found out that it is another hoax, pods of *B. variegata* superposed on a sterile leafy branch which 'probably belongs to *B . scandens* L'.

11. Later Isonym When the same name based on the same type, has been published independently at different times by different authors, in which the earliest of these, the isonym has the nomenclatural status. Whereas later isonyms published are disregarded. For e.g. Baker (1892) and Christensen (1905) independently published the name *Alsophila kalbreyeri* as a substitute of *A. podophylla* Baker (1891), hence the name *A. kalbreyeri*, being a later isonym, is to be discarded.

12. **Nomina rejicienda** Names that are widely and persistently used for a taxon which does not include its original type (nomenclatural type) and also latin diagnosis should be rejected and placed under the list of nomina rejicienda(Art.69).

In general

- Illegitimate epithets are not taken into consideration for purpose of priority except in the rejection of later homonym.
- A specific epithet must not be considered as illegitimate merely because it was originally published under an illegitimate generic name, but it must be taken into consideration for the purpose of priority, if the epithet and corresponding combination are in accordance with the rules of botanical nomenclature. The same rule is also applied for an infraspecific epithet published under an illegitimate name of a species.
- The original spelling of a name must be retained except in case of Typographic or Unintentional Orthographic Error.

For e.g. *Rosa pisardii* Carr. is corrected as *Rosa pisardi*, and *Pinus griffithi* is corrected as *Pinus griffithii*.

The aim of the present Code is to secure uniform and stable system of plant nomenclature and as a whole it is an excellent Code. It is the extract of the intense study made by the best brains in the botanical fraternity, who for nearly a century has been analysing the hurdles connected with naming of plants, so as to bring the botanical nomenclatural system to satisfactory in all respects (Santapau, 1965).

15

ORTHOGRAPHY OF NAMES

Keywords Orthographic error, typographic error, orthographic variants, examples.

Articles 73-75 of the Code deal with orthography of plant names. According to the Code, "the original spelling of a plant name is to be retained unless it is orthographically wrong or can be shown to be a misprint when published.

The orthographic names may be of the following types:

1. **Retention of the original spelling** The generic name *Amaranthus* L. 1753 was spelled by Linnaeus and the spelling of the name is not to be altered to *Amarantus,* as found in most of the subsequent publications of the Indian text.
2. **Orthographic error** *Gluta benghas* L. (1753) is an orthographic error for *Gluta renghas* L. when Linnaeus published this species; he used the vernacular name 'Renghas' as a specific epithet of this species. Hence the binomial should be cited correctly as *Gluta renghas* L. as was done by Engler (1883).
3. **Typographic error** The species *Globba brachycarpa* Baker (1890) is a typographic error for *Globba trachycarpa*

Baker. Similarly *Hetaeria alba* Ridley (1896) is also an example of typographic error for *Hetaeria alta* Ridley.

4. **Orthographic variants** When two or more generic names are more or less similar leading to confusion, these names are treated as variants, which are homonym when they are based on different types. For e.g. *Pleuropetalum* and *Pleuripetalum* are considered as orthographic variants, in which only one name can be used without serious risk of confusion.

Whereas *Rubia (Rubiaceae)* and *Rubus* (Rosaceae) are two different generic names which do not lead to any confusion, hence, they are not treated as orthographic variants.

Likewise, the specific epithets – *sinensis* and *chinensis*, *ceylonica* and *zeylanica*, *macrocarpon* and *macrocarpum*, are treated as orthographic variants as they are likely to be confused when applied to related species within the same genus and also to infraspecific epithets within the same species. The names ought to be preserved when the changes made in orthography by the earlier authors adopted personal, geography, or vernacular names in nomenclature are intentional latinization, except for the wrong use of the terminations like i, ii, ae, iae, anus and ianus is treated as an orthographic error, which is to be corrected. For e.g., *Rosa pissarti* Carriere is an e.g. for typographic error for *Rosa pissardi*, which is in turn treated as an orthographic error for *Rosa pissardii*.

The use of hyphen after a compounding form in an epithet is treated as an orthographic error and it is to be corrected, but the same rule is not applied to the names of genera or of higher ranks. However, a hyphen is permitted in an epithet after a word which could stand independently.

For e.g.

1. *Hibiscus rosa- sinensis.*
2. *Acer pseudoplatanus* L. but not *Acer pseudo-platanus*, in which the epithet *pseudo* cannot stand as a complete word independently and this rule is applied only to epithets.
3. *Veronica anagallis- aquatica*, in which *anagallis* can stand as a complete word independently.

16

NOMENCLATURE TERMINOLOGY

Keywords Definition, Latin names, legitimate name, illegitimate name, alternative name, correct name, rejected name, naked name, latin diagnosis, catalogue, tautonym, conserved name, confused name, superfluous name, basionym, autonym, synonym, type specimen, homonym, new combination, parenthetical authority, orthographic name, latin diagnosis, examples.

In plant taxonomy all the names are treated as Latin names. The names are treated according to the principle of the Code of Botanical nomenclature. The principle V states that "Scientific names of taxonomic groups are treated as Latin regardless of their derivation". According to the rules of nomenclature, the name of a taxon has no status unless it is validly published. Such names may be legitimate or illegitimate.

The different names and their definitions used in Code are as follows:

1. **Legitimate name** A name or an epithet published in accordance with all the rules of the Code is referred to as a valid name or legitimate name (Art.32-45). It may be correct or incorrect according to different concept of the taxa.

   e.g., *Annona squamosa* L.

2. **Illegitimate name** A name which is not validly published or not in accordance with the rules of ICBN is called illegitimate name. An illegitimate name must be rejected under the Code. e.g., Wallich's Catalogue, superfluous names and tautonyms etc. However, an illegitimate name may become legitimate if it is conserved as in the case of the family *Ebenaceae* Giirke, nom. cons. This family name is based on the stem (type genus) of an illegitimate generic name *Ebenus* O, Ktze., which is illegitimate unless conserved (Art.18.3). Hence, this family name is (*Ebenaceae*) legitimate as it is conserved.

3. **Alternative name** Each taxon of the rank of family and below can bear only one correct name, but the following families have alternative names , which are permitted by ICBN (Art.18,Note 3).

Name under usage	**Name according to rule**
Cruciferae	Brassicaceae
Compositae	Asteraceae
Gramineae	Poaceae
Labiatae	Lamiaceae
Leguminosae	Fabaceae
Palmae	Arecaceae
Umbelliferae	Apiaceae
Guttiferae	Cluciaceae

4. **Correct name** A legitimate name that is adopted under the rules for a taxon with a particular circumscription, position and rank is a correct name (Principle IV). A legitimate name may be correct or incorrect according to different concepts of the taxa.

   For e.g., the generic name *Hibiscus* Linn.(1753) is a legitimate name because it was published in accordance with the rules of the Code. Similarly the same is true of the generic name *Abelmoschus* Medicus (1787). Both the genera are correct when they are considered separately. But in 1824 Augustin de Candolle united *Abelmoschus* and *Hibiscus* to form a single genus, and when this treatment was accepted, as followed in most of the earlier Indian Floras, *Hibiscus* L. was considered as the only correct name with this particular circumscription. But K. Schumann (1895) and several recent Monographers re-established the genus *Abelmoschus* Medicus. Hence, the legitimate generic name *Abelmoschus* may therefore be correct or incorrect according to different perceptions of the taxonomists of the taxa concerned.

5. **Rejected name** A name must be rejected if it has been widely and persistently used for a taxon not including its type (type specimen), or nomenclaturally superfluous, or a later homonym or a tautonym. Generally all the illegitimate names are the rejected names, which must not be taken into consideration for the purpose of priority, except in the rejection of later homonym.

6. **Naked name** A name which is published without description or latin diagnosis or reference is known as naked name or nomen nudum (nom. nud). This name is not valid according to the code and is to be rejected.

   For example most of the names published without any description in Roxburgh's Hortus Bengalensis (1814) and Wallich's Catalogue are naked names. These names can be later on validated by other author(s) by giving description or Latin diagnosis for the species.

7. **Tautonym** A binary name in which, the specific epithet exactly repeats the generic name is called a tautonym. Such name is illegitimate and not admissible in ICBN (Art.23). The following are the examples of tautonyms:

   i. *Linaria linaria*

   ii. *Lycopersicon lycopersicon*

   iii. *Malus malus*

8. **Conserved name** In order to avoid disadvantageous changes in the nomenclature, certain names of genera and families are conserved by the Code due to their long usage in the literature and such names are called conserved names (Art.14) Conserved names are abbreviated as nom. cons.

   For example the genus *Tectona* Linn. (1781) (nom. cons.) is conserved against its earlier name (*Theka* Adans. 1763) and the eight family names ending with **'ae'** (*Compositae*) are also conserved by the Code .

9. **Confused name** A name based on a type consisting of discordant element is termed as a confused name and it should be rejected.

   Polygala chinensis (nom. confus.) is an example of a confused name.

10. **Superfluous name** A name applied to a taxon for which another name is already available is a superfluous name, and it should be rejected.

    For example *Cyperus alulatus* Kern. is a correct name. Whereas *Cyperus rectangularis* (Kuk) Bennet is a superfluous name, because *Cyperus alulatus* Kern. is already available. Hence, the superfluous name *Cyperus rectangularis* should be rejected.

11. Basionym When a new name or an epithet required by a change in the position or rank of a taxon , is taken from a previous name for the taxon, is called the basionym of the new name. It is also called epithet bringing name.

For example, when *Lycopodium pallescens* Presl. was transferred to the genus *Selaginella* by Spring, it becomes *Selaginella pallescens* (Presl.) Spring. In the above citation, *Lycopodium pallescens* is the basionym, on which the name is based. When the genus is transferred from one genus to another genus, the specific epithet if it is valid should be retained in the new combination, (*Selaginella pallescens*) unless there is some nomenclature barrier that exist, and the author of the basionym should also be cited in the parenthesis. The new combination is also an example of double citation.

Draba obscura Dunn. Kew Bull. 383. 1924 is the basionym for *Lignariella obscura* (Dunn.) Jafri, Candollea 134. 1957.

Brassica campestris L. is the basionym for *Brassica rapa* subsp. *campestris* (L.) Clapham *et al.*

12. Autonym The name which is automatically established for a taxon, merely as a consequence of the proposal of another taxon is known as an autonym.

For e.g., the proposal of a subspecies or variety that does not contain the type of the species automatically establishes a subsps or variety that does contain the type. These names are established below family and genus level and are not taken into consideration for the purpose of priority (Art.26).

For example Keller divided the species *Rosa gallica* into variety level. As a result the variety that does not contain the type of the species automatically establishes a variety that does contain the type. Hence, the final epithet of an autonym at the infraspecific level is that of the specific name unaltered, but without citation of author's name. e.g.. *Rosa gallica* Linn. var. *gallica* – autonym. He named the other variety as *Rosa gallica* var. *eriostyla* Keller. (The autonyms are always written without any author's citation).

Clapham (1962) reduced *Brassica campestris* L. to subspecies level in *Brassica rapa* L., so it was established as an autonym e.g., *Brassica rapa* L. subsp. *rapa* without any citation of author and *Brassica campestris* became *B. rapa* subsp.*campestris* (L.) Clapham.

13. **Ambiguous name** It is a name used in different senses, so that it has become a long source of persistent error. Ambiguous names should be rejected (Art.69).

    For example, the name *Polygala chinensis* L. has been applied to two distinct species. One is found almost throughout India which is *Polygala arvensis* Willd. and the other found in South East Asia and in Malaysia is *Polygala chinensis* L. Therefore the name *Polygala chinensis* L. being a source of confusion, as it has been applied to two different species equally, Burt pointed out that the *Polygala* species found in India is not the species that was described by Linnaeus. Hence, Burt has rejected *Polygala chinensis* L. and rightly adopted the next earliest available correct name *Polygala glomerata* Lour.

14. **Synonym** The different names used for a single taxon or a taxonomic group are called synonyms. The synonyms usually follow a valid name and are italicized and are to be rejected.

There are two kinds of synonyms recognized by the Code such as:

i. nomenclatural synonym
   and
ii. taxonomic synonym.

i. **Nomenclatural synonym** Names based on the same type are called nomenclatural synonyms, or homotypic synonyms. For example *Justicia santapaui* Ben (1970, new name), *J. andersonii* Raman (1976), *J. montana* (Nees) Wall. ex T.Anderson (1805) and *Hemichoriste montana* Nees (1832) are all based on the same type and there is no question of their identity. Hence all the names are considered as nomenclatural synonyms. The correct name of the plant is *Justicia santapaui*, which is named in honour of the well known systematic botanist Fr.Santapau.

ii. **Taxonomic synonym** These are the names based on different types, belong to the same taxon usually known as taxonomic synonyms or heterotypic synonyms.

For example 1. *Hibiscus vitifolius* L. (1753) and *H. obtusifolius* Willd (1801) are the names based on two different types. An examination of both the type specimens revealed identical. This, of course is entirely based on a taxonomic judgement. Hence, *Hibiscus vitifolius* L. is the correct name (priority wise) and *H. obtusifolius* Willd is a taxonomic synonym of *H. vitifolius*.

2. *Boerhaavia erecta* Linn. (1753) is described from tropical America whereas *B. punarnava* Saha & Krish (1961) is identified from South India. But both the names are based on different types. In 1968, when Nair analysed, he came to a conclusion that, both *B. punarnava* and *B. erecta* are referring to a same taxon and merged *B. punarnava* under *B. erecta*. Therefore, both the names are considered as taxonomic synonyms.

15. Homonym A name which is spelt exactly like a validly published name for a taxon of the same rank based on a different type (Art.64) called a homonym. In this case the later homonym is illegitimate hence it is replaced by a new name.

For example, *Jatropha heterophylla* Heyne ex Hook.f. (1887) is a later homonym of *Jatropha heterophylla* Steudel (1840). In this case, the later homonym i.e., *J. heterophylla* Heyne ex Hook.f. (1887) is an illegitimate name and it should be rejected. But it is replaced by a new name called *Jatropha heynei* Bal. (Bull. Bot. Surv. India 3:41.1961) in honour of the Botanist Heyne and accepted as a legitimate name

Distemon Wedd. 1857 of *Urticaceae* is a later homonym of *Distemon* Bouche 1844 of *Cannaceae*. Babu and Henry in 1970 validated and proposed a new name *Neodistemon* Babu & Henry for *Distemon* Wedd. (1857).

16. **New Combination (Comb. nov.)** A new name formed from a previously published name or epithet and typified by the type of basionym is designated as comb. nov. or new combination. When a taxon of lower rank than genus is transferred to another genus or species, with or without alteration of rank, based on a taxonomic judgement, the original name or epithet when legitimate,(basionym) is retained and the author of the basionym is cited in parenthesis (Bracket) followed by the name of the author who effected the alteration.

    For example *Arabidopsis verna* var. *hebecarpa* N.Busch. was transferred to *Drabopsis* by Naqshi & Javeid, it becomes a new combination as *D. verna* var. *hebecarpa* (N. Busch) Naqshi & Javeid, Comb. nov.

17. **Nomen novum (Nom. nov.)** A substitute name is to replace an earlier name and typified by the type of earlier name is known as nomen novum or new name. A new name is chosen only when no legitimate name is available.

For example *Emblica officinalis* Gaertn. (1791) is a new name for *Phyllanthus emblica* L. when *Phyllanthus emblica* L. (1753) was transferred to the genus *Emblica*, it became *Emblica emblica,* an example of a tautonym, which was inadmissible to the Code. Therefore Gaertner proposed a new specific epithet *officinalis* and called the plant by the new name (nom. nov.) *Emblica officinalis* Gaertn. (1791).

18. The Authority The name of a person(s) written after the binomial is called the authority of the name. Authority is given when a taxon or family is originally described, named and published it by a person(s).

For example *Annona squamosa* was first named, described and published by Linnaeus, hence, he became the authority for that (plant) name. Hence, it is written as *Annona squamosa* Linn. The name of the authority, unless very short, should be abbreviated.

19. The Parenthetical Authority When the rank of a taxon or taxonomic group is changed or a species is transferred from one genus to another or similar changes are made means, the name of the original author is placed in parenthesis (Bracket) after the binomial and then followed by the name of the person(s) who is responsible for the change or transfer.

For example, *Medicago polymorpha* L. var. *orbicularis* was first named by Linnaeus, when raised to the rank of a species by Allioni, it becomes *Medicago orbicularis* (Linn) All. It is also an example of double citation.

20. Orthographic name The original spelling of a plant name is to be retained unless it was definitely wrong orthographically or can be shown to be a misprint when published.

For example, the generic name *Clerodendrum* L. (1754) so spelled by Linnaeus is not to be altered to *Clerodendron* as found in most of the subsequent publications.

21. Latin Diagnosis The diagnosis of a species or an infraspecific taxon may be very short with a statement of one or more characters not possessed by the other species of the same genus. It is recommended by the Code that authors publishing names of new taxa of recent plants should give a full description in Latin language, in addition to the diagnosis.

17

NOMENCLATURE OF HYBRIDS AND CULTIVATED PLANTS

Keywords Introduction, hybrids, nothotaxon, nothogenera, nothospecies, hybrid formula, condensed formula, graft hybrids, cultivars, examples.

A. NAME OF HYBRIDS

Introduction An offspring obtained from two different genera or species of plant is called a hybrid. In botanical nomenclature, the hybrid released is given a 'hybrid name' which is a special kind of botanical name. When a hybrid is formed naturally it may get a normal botanical name.

ICBN has separate rules for hybrids and Appendix I of the Code, based on twelve articles (H.1-H.12) pertaining to the formation of name of hybrids. The provisions for hybrids (Appendix I) were thoroughly overhauled at Sydney (Voss *et al.*, 1983). However, there were no significant changes except for the adoption of the word 'nothotaxon', to refer to a taxon, at any rank, of hybrid origin. Therefore the ranks of taxon such as nothogenera, and nothospecies available for different hybrids and all these taxa are designated as nothotaxa. The foregoing account deals only with the provisions of the Code relevant to wild flowering plants.

Hybridity is indicated by the use of the multiplication sign '×' or formula or a new name or by the addition of the prefix 'notho'- to the term denoting the rank of the taxon called nothotaxa, to indicate their hybrid nature and their putative parents (known parentage).

A hybrid named between taxa is indicated by placing the multiplication sign'×' simply between the names of both the parents and the whole expression is then called a 'hybrid formula'.

For e.g.,

1. *Agrostis* L. × *Polypogon* Desf.
2. Agrostis stolonifera L. × Polypogon monspeliensis (L.) Desf.
3. *Salix aurita* L. × *S. caprea* L.

It is recommended that the names or epithets in a formula to be placed in alphabetical order. The direction of a cross may be indicated by including the sexual symbols **({♀} : female; {♂} : male)** in the formula, or by placing the female parent first. If a non-alphabetical sequence is used, its basis should be clearly indicated.

A hybrid may be 'interspecific' (between two species of the same genus) or 'intergeneric hybrid' (between two species of two different genera).

It is important to note that the hybrid derived between two or more taxa may receive a name and for nomenclatural purpose, a binomial for an interspecific hybrid of a taxon is denoted by placing the multiplication sign '×' in the name of a nothotaxon against the initial letter of the name or epithet. However, if mathematical symbol is not available and the letter 'x' in lower case may be used leaving a single letter space before the specific epithet in the name of an interspecific hybrid (Hybrid formula may be added within the parentheses if the parents are established**)**, or multiplication sign '×' is placed before the name of an intergeneric hybrid or by prefixing the term 'notho'- to the term denoting the rank of the taxon such as 'nothospecies' for interspecific hybrid and 'nothogenus' for intergeneric hybrid.

For e.g.,

1. *Salix × capreola* Andersson (1867)
2. *Salix × capreola* (*S. aurita × S. caprea*)
3. *Polypodium vulgare* nothosubsp. *montoniae* (Rothm.) Schidley.
4. *Mentha × smithiana* R.A. Graham (1949)
5. ×*Agrohordeum* (*Agropyron × Hordeum*)

A nothotaxon cannot be designated unless one of the parent is known or can be postulated. However, taxa which are believed to be of hybrid origin need not be designated as nothotaxa.

The multiplication sign '×'and the prefix 'notho'- are deleted for the purpose of homonyms and synonyms. For e.g., × *Hordelymus* Bachteev & Darevsk. (1950) (= *Elymus* L. × *Hordeum* L.) is a later homonym of *Hordelymus* (K.Jess.) K.Jess. (1885).

The appropriate rank of a nothotaxon remains the same in its rank as that of known parents; however, if parent taxa are of unequal rank, the appropriate rank of the nothotaxon becomes the lowest of these rank.

Sometimes it occurs when a taxon designated by a name in a rank is inappropriate to its hybrid formula, the name may become incorrect to that particular hybrid formula, however, it may become correct later on when the mistake is rectified by the author.

1. **Name of Interspecific Hybrid** The hybrid formed between two species of the same genus is called interspecific hybrid or nothospecies and the name is usually written in two different ways:

   a. The name of both the parents are simply connected by a multiplication sign, '×' and the whole expression is then called a 'hybrid formula'.

      For e.g.,

      1. *Salix aurita* × *S. caprea*
      2. *Verbascum lychnites* × *V. nigrum.*
      3. *Mentha aquatica* × *M. arvensis*
      4. *Digitalis purpurea* × *D. lutea*

   b. Hybrid between the species of the same genus may be provided with distinct name (new epithet) of their own with '×' sign. The generic name and the new epithet are connected by a multiplication sign (×), as cited below:

      For e.g.,

      i. *Salix* × *capreola* (*Salix aurita* × *S. caprea*)
      ii. *Verbascum* × *nigralychnites* (*Verbascum lychnites* × *V. nigrum*)
      iii. *Rosa* × *odorata* (*Rosa chinensis* × *R . gigantea*)

Such hybrids may be given altogether a new name when parents are not definitely known. Thus, *Musa* × *paradisiaca* L. indicates that this species has a hybrid origin.

The epithet in the name of a nothospecies is called a 'collective epithet'. There can only be one correct name to a particular hybrid formula, which is the earliest legitimate name (Art.6.3) in the appropriate rank (Art.H.5) and other names to which the same hybrid formula applied are synonyms.

For e.g., The names *Oenothera* × *wienii* Renner ex Rostanski (1977) and *Oenothera* × *hoelscheri* Renner ex Rostanski (1968) both apply to the hybrid *O. rubricaulis* Kleb. × *O. depressa* Greene,the types of the two nothospecific names differ from each other by a whole gene-complex; even then the later name is treated as a synonym of the earlier.

The variants of interspecific hybrids are named 'nothosubspecies', and 'nothovarieties', e.g.,1. *Polypodium vulgare* nothosubsp. *montoniae* (Rothm.) Schidley, and *Salix rubens* nothovar. *basfordiana*. When a name of a new nothotaxon is published at the rank of species or below, the author should provide the available information for its taxonomic identity and at the lower rank the type of the name should be given for the known parent plants.

The name of nothotaxa at the rank of species or below must conform with the provisions of the Code applicable to the same rank and since the name of a nothotaxon at the rank of species or below has a type, statements of parentage play a secondary part in determining the application of the name.

2. **Name of Intergeneric Hybrids** The hybrids derived between two species of two different genera are termed intergeneric hybrids or nothogenera. A nothogeneric name of a bigeneric or two generic hybrid is formed with a distinct generic name by combining both the parental generic names into a single word and the name is formed by using the first part of or whole of one parental genus

and last part or whole of other genus, but not whole of both the genera, connected by a vowel. This intergeneric name is so-named as "nothogeneric name" which is known as a "condensed formula". The combined generic name is preceded by a multiplication sign '×' (since it is a hybrid between species of two different genera). That is, the multiplication sign is placed before the combined generic name of the hybrid as given below:

For e.g.,

a. × *Heucherella* (from condensed names of *Heuchera* and *Tiarella*).

b. × *Aspleniophyllitis* (from condensed names of *Asplenium* × *Phyllitis*).

c. × *Triticale* (from *Triticum* × *Secale* or *Triticosecale* from *Triticum* × *Secale*).

d. × *Pyronia* (from *Pyrus* × *Cynodia*).

e. × *Agropogon* (from Agrostis × *Polypogon*).

f. × *Agrohordeum* (from *Agropyron* × *Hordeum*).

g. × *Cooperanthes* (from *Cooperia* × *Zephyranthes*).

h. ×*Cupressocyparis* (from *Chamaecyparis* ×*Cupressus*)

However, a nothogeneric name of an intergeneric hybrid if derived between four or more genera, is formed from the name of a person, to which, is added the termination- '**ara**', and such names must not exceed eight syllables. These names are also regarded as condensed formula.

For example,

× *Potinara* Charlesworth & Co.(1922) (= *Brassavola*

× *Cattleya* × *Laelia* × *Sophronitis*)

In case where a nothogeneric name of a trigeneric hybrid developed is either

i. a name formed like that of a nothogenus derived from four or more genera i.e., from a name of a person to which is added the termination- **ara**, for example, ×*Wilsonara* Charlesworth & Co. (= *Cochlioda* × *Odontoglossum* × *Oncidium*). When a nothogeneric name is formed from the name of a person by adding the termination- **ara**, that person should preferably be a collector, grower, or student of the group.

or

ii. a condensed formula in which three parental generic names are combined into a single word not exceeding eight syllables, using the whole or first part of one, followed by the whole or any part of another, followed by the whole or last part of the third but not the whole of all the three and with the use of one or two connecting vowels as given below:

For e.g.,

1. ×*Vascostylis* Takakura (= *Ascocentrum* × *Rhynchostylis* × *Vanda*)
2. × *Sophrolaeliocattleya* Hurst. (= *Cattleya* × *Laelia* × *Sophronitis*).

The name of a nothotaxon, which forms a hybrid between subdivisions of a genus is a condensed formula formed in a similar way as a nothogeneric name with the name of the genus.

For e.g., *Ptilostemon* nothosect. *Platon* Greuter (1973), comprising hybrid between *Ptilostemon* sect. *Platyrhaphium* Greuter and *Ptilostemon* Cass. sect. *Ptilostemon*.

When the name of a nothotaxon is a condensed formula, the parental names used in its formation must be those which are correct for the particular circumscription, position and rank accepted for the parental taxa. These names should be published effectively with a statement of the names of the parent genera or subdivisions of genera, however, no diagnosis or description is necessary whether it is Latin or in any other language.

A binomial for the intergeneric hybrid may similarly be written as under:

× *Agropogon lutosus* (*Agrostis stolonifera* × *Polypogon monspeliensis*).

Since the names of nothogenera and nothotaxa with the rank of a subdivision of a genus are condensed formulae or treated as such, ,as they do not have types.

It is important to note that a binomial for an interspecific hybrid has a multiplication sign '×' before the specific epithet, whereas in an intergeneric hybrid it is placed before the generic name.

3. **Graft hybrids** The graft hybrid (or graft chimeras) formed between two species is named with a plus sign (+) placed between the two grafted species as, for example, *Rosa webbiana* + *R. floribunda*, or with the + sign placed in front of the name of graft hybrid as for e.g., + *Cratagomespilus dardarii* or *Cratageus monogyna* + *Mespilus germanica*.

B. Nomenclature of cultivated plants

A cultivar is an assemblage of cultivated plants, which can be very well distinguished by its characters. A cultivar has the capacity to retain its distinguishing characters when reproduced sexually, or asexually. It may be derived from a clone, a group of identical clones inbreds or lines, or a series of cross- fertilized individuals and so on, but it should be maintained by cultivation.

As these plants do not have a structure of natural population, they have to be propagated and maintained as artificial population by man. This is the reason why the usual botanical hierarchy practised for other plants cannot be applied to these cultivars and therefore separate rules have been proposed for these cultivars. Accordingly the names of cultivated plants are governed by International Code of Nomenclature for Cultivated Plants (ICNCP), which was published by Trehane *et al.*, (1995). Most of the rules of ICNCP are taken from ICBN with an addition of a separate rank 'cultivar', abbreviated as 'cv' for cultivated varieties.

Cultivar is an internationally recognized term for the category of distinct cultivated plants, in preference of the term 'variety', is used in Britain. As the term 'varietas' constitutes a different botanical category, should not be confused with variety of cultivars.

Cultivar names are written with capital initial letter and need not be italicized or in Latin, but rather a common name. The cultivar name is either preceded by the abbreviation cv. (cultivar) as in *Citrullus lanatus* cv. Sugar Baby, *Cucurbita pepo* cv. Table Queen, *Rosa floribunda* cv. Blessings, and *Primula vulgaris* cv. Lilac Queen, or simply written within single quotation marks, for e.g., *Rosa floribunda* 'Blessings'. Cultivars may also be named directly under a common name, as for example, Hybrid Tea Rose 'Red Lion' or directly under a genus e.g., *Hosta* 'Decorata' or under a hybrid e.g., *Rosa* × *paulii* 'Rosea'. The correct nothogeneric name for plant derived from the *Triticum* × *Secale* crosses is × *Triticosecale* Wittmack ex A. Camus. At the species level no correct name is available for the commonest crop Triticales. It is recommended that the crop Triticales be named by appending the cultivar name to the nothogeneric name for e.g., × *Triticosecale* 'Newton'.

Since 1.1.1959, new cultivar name should have a description published in any language and the name must not be the same as that of botanical or common name of a genus or a species. Thus cultivar name 'Rose', 'Onion' etc. are not permitted as cultivar name. It is also recommended by the Code that cultivar name must be properly registered with 'registering authorities' for various plant groups to prevent duplication or misuse of cultivar names. Registering authorities exist separately for Roses, Orchids and several other groups or genera.

Cultivar name remain unchanged when their botanical name change. For example, corn flower "Blue Diadem" remains the same whether the species is called botanically as *Centaurea cyanus* or *Cyanus segetum*.

18

NOMINA CONSERVANDA

Keywords Definition, Botanical congress, principle, priority, conserved names, appendices, nomina generica conservanda, nomina familiarum conservanda, nomina specifica conservanda, examples.

A number of names of families, genera and species are legitimated by 'conservation' by the Code due to their very long usage in the botanical literature and also in the establishment of nomenclatural stability of names.

The concept of Nomina Conservanda was introduced at the Third International Botanical Congress held in Vienna in June 1906, which was the first Botanical Congress that functioned International in character. This system was introduced for conservation of certain genera and families just to avoid disadvantageous changes in the names of families and genera by the strict application of the principle of priority.

When a taxon is accepted by the committee for conservation, the abbreviation nom.cons. (nomen conservandum) is added to the citation. A special committee was appointed by the Montreal Code (1961) at the Ninth International Botanical Congress to study the question of conservation of family names, and an Appendix II with Nomina Familiarum Conservanda was introduced in the Code. In the Edinburgh Code (1966) and Seattle Code (1972) some amendments in the list of Nomina Conservanda were made and a list of Nomina Conservanda was introduced. Accordingly an amendment in the Nomina Familiarum Conservanda was also made and the family names of recently discovered Gymnosperms were also added in addition to the families of Angiosperms. In the selected list of Nomina Familiarum Conservanda, a few names are found with a somewhat different spelling from the long-established names. For e.g., *Cannabaceae* for *Cannabinaceae, Capparaceae* for *Capparidaceae*, *Haloragaceae* for *Haloragidaceae*, *Melastomataceae* for *Melastomaceae* etc.

A. **Nomina Familiarum Conservanda** The list of conserved family names (Nomina Familiarum Conservanda) was included in Appendix II of the earlier Codes, but in recent Codes , Nomina Familiarum Conservanda is placed in Appendix IIA and IIB, where Appendix IIA with conserved family names of Algarum, Fungorum, Pteridophytorum Et Fossilium Et Rejicienda and Appendix IIB includes the rest of Nomina Familiarum Bryophytorum Et Spermatophytorum Conservanda.

Many of the names included in the list are not in dispute and hence not in need of conservation, but their inclusion in the list is useful as a ready reference work giving correct names of families together with the type genus for each family. The starting date for Nomina familiarum conservanda is AL de Jussieu, *Genera Plantarum* (1789) and earlier usage of listed names are to be rejected.

Van Steenis (1954) suggested the following as guiding principles in the matter of conservation:

i. That the spirit of the Code is to promote stability of nomenclature, and that this supersedes the principle of priority, where possible, reasonable and desirable.

ii. That conservation of generic name is intended to stabilize nomenclature, irrespective of size or economic importance of taxa.

iii. That consequently those names must be conserved which are the longest in general use, and that obsolete names, whatever the scientific merits of their authors, should not replace names which are in current usage.

B. **Nomina Generica Conservanda** For the first time in the history of botanical nomenclature, the establishment of conserved generic names (Nomina Generica Conservanda) was made at the Vienna Code (1906), in which the generic names have wide usage should be considered valid and conserved over the name of the same individual published prior to it. Therefore a list of generic names of plants that are having wide usage over the earlier ones but less known was introduced as conserved names under the list Nomina Generica Conservanda. But the American Code (1907) refused to accept the principle of Nomina Generica Conservanda.

In the Sixth International Botanical Congress held in Amsterdam (1947), a list of Nomina Generica Conservanda was again included in the Code. Later in Paris Code (1956), that is at the Eighth IBC held also the Nomina Generica Conservanda et Rejicienda was amended and supplemented and listed in Appendix III of the earlier code but in recent codes it is placed in Appendix IIIA. The present list of nomina generica conservanda is based on Nomina generica conservanda et rejicienda Spermatophytorum by HW Rickett and FA Stafleu (1959-61) published in Taxon. When a generic name is conserved, its exact circumscription must be indicated and this point needs a careful study.

The Appendix IIIA of the current Code (Vienna Code, 2005) provides a list of conserved and rejected names of genera (Nomina Generica Conservanda Et Rejicienda). For example, the generic name *Desmodium* Desv. (1813) is conserved against the earlier names *Meibomia* Heist. ex Fabricius (1759) and *Pleurolobus* J. Saint- Hilaire (1812), due to its very long usage in Botanical publication.

The obsolete generic names (*Meibomia* and *Pleurolobus*) though earlier, were rejected. The names in such lists are conserved against all earlier homonyms, and nomenclatural synonyms (different names based on the same type); and all names based on different types mentioned in the list.

Any proposal for conservation of names has to be submitted to the General Committee with a clear statement of the case both far and against its conservation. This will be referred for the examination to the Committees for the various taxonomic groups. The proposal of conservation have to be approved by the General Committee of the concerned taxonomic group and then by the next International Botanical Congress by voting system.

When a conserved name competes with one or more other names based on different types and against which it is not expressly conserved, the earliest of the competing name is adopted against the conserved one. For e.g., *Nasturtium* R.Br.(1812) was conserved only in the restricted sense for a monotypic genus based on *N. officinale* R.Br. If it is reunited with *Rorippa* Scop. (1760), it must bear the prior name *Rorippa*.

C. **Nomina Specifica Conservanda** Many attempts have been made in the past to include a list of 'Nomina Specifica Conservanda' in the Code and all these attempts have proved unsuccessful due to various reasons for many years. Therefore conservation of specific names is not legislated in the 'Code' until the establishment of

Sydney Code (1983). Strict application of the principle of priority has caused many changes in the names of species.

The identity of the plant may have been more clearly understood, this often means a change in specific name for e.g., Santapau (1960) has shown that the common species of *Entada* of Bombay and Western ghats of India is not *E. scandens* Benth. nor *E. Phaseoloides* Merr. but *E. pursaetha* Dc. The latter species is distributed from Tropical and Central Africa eastwards to the Western part of India and perhaps to the south, whereas *E. scandens* comes from East Bengal, Assam etc., down to the Mollucas. It is obvious that the change in specific name is justified. Again Pampanini (1926) showed that none of the Indian species of *Artemisia* corresponds to the European *A. vulgaris* L., even though our plants are so called in our floras, Pampanini changed the name to *A. nilagirica* Pamp. and here again the change of name is fully justified.

In some instance the oldest specific epithet was not used for the plant, and this is prescribed in the Code. This is a constant source of annoyance to the Botanists, Horticulturists, Foresters and so on. Merrill (1912, 1916), however, often expressed the opinion that such changes will be necessary only as long as there remains some little known botanical works to explore, the number of such changes is becoming rather scanty at present, so that one may foresee the time when no further change will be necessary on this account. It is precisely against such changes that conservation of specific names has been demanded.

During the past, some names have been changed on the plea that they were inappropriate, for e.g., the plant called *Vigna capensis*, reported from India or Ceylon and not found in Cape, others with specific names indicating a colour which is not found in the plant. The Code rules out such changes. A legitimate name must not be rejected simply because it is inappropriate or disagreeable.

The present list of Nomina Specifica Conservanda Et Rejicienda is listed in Appendix IIIB. Though there has been hot discussions for years since Cambridge Congress of 1935. No general agreement has been made for conservation of specific names. Lanjouw (1963) also suggested a solution for the problem to conserve only important plants. Here it is very difficult to decide what are the important plants? Fosberg (1964) strongly objected to Nomina specifica conservanda on the plea that such a list would render the Code almost unusable. De Wit (1964) states that many plants which were important economically have now lost their economic importance wholly or partially. An attempt to select a list of Nomina specifica conservanda was thwarted by an overwhelming vote at the Sixth International Botanical Congress held in Amsterdam (1947). Therefore the Code rejected the establishment of a list of Nomina Specifica Conservanda by 208 to 61 votes, and in the same way the Stockholm Code (1952) also rejected the principle of Nomina Specifica Conservanda by a ballot of 320 votes to 40 was adopted.

The adoption of a list of Nomina Specifica Rejicienda which also had the same stabilising effect as introducing a Nomina Specifica Conservanda. In the Montreal Code (1961) also Nomina Specifica Conservanda was not accepted but it was agreed that an attempt may should be made to assess the importance of the problem before the rule was changed. Later in Sydney Code (1983), conservation of names of species was accepted but was not included in the Code. The Nomina Specifica Conservanda was introduced for the first time in Berlin Code (1988) and listed in the Appendix IIIB as Nomina Specifica Conservanda Et Rejicienda. At this Congress two species name such as *Lycopersicon esculentum* P.Miller and *Triticum aestivum* Linn. have been conserved against the rule of priority, as these names are widely used and any change may create confusion.

19

PUBLICATION OF A NEW SPECIES

Keywords Rules, generic name, recommendations, binary name, latin terminations, latin diagnosis, specific epithet, binary combination, intergeneric hybrids, interspecific hybrids, tautonym, nomenclatural type, parasites, examples.

A name of a taxon has no status in the Code unless it is validly published following the rules of botanical nomenclature. The conditions and dates of Effective and Valid publication of names of taxa have been dealt in Articles 29-45 of the chapter IV.

When an author(s) describes a new genus, the name of a genus must be accompanied by a detailed description or diagnosis of the genus followed by description or reference to an earlier validly published description of the type species of that genus if exist. Monotypic new genus based on a new species is validated by the provision of a combined generic and specific description or diagnosis.

New genus It is recommended by the ICBN that, during the formation of any generic name one must keep in mind that the following rules to be followed strictly:

1. a. Generic name is a substantive noun in the singular number treated as such and is always written with a capital initial letter.

b. The generic name may be taken from any source whatever and even be composed arbitrarily (Art. 20.1).

c. The generic name should not coincide with a technical term currently used in morphology, unless it was published before 1.1.1912 and accompanied by a specific name. For e.g., the generic name *Radicula* Hill (1756) coincide with a morphological technical term radicle and it was also published without binary name. Hence the generic name *Radicula* was rejected. Whereas other generic name *Tuber* Wigg (1780) was published with specific epithet as *Tuber gulosorum* Wigg. Hence it was accepted as a validly published name even if the generic name denotes the morphological term.

d. The name of a genus must be in one word, and if it is formed of two words, these words must be joined together by a hyphen (Art.20).

e. The generic names must have a Latin termination and names not readily adaptable to Latin should be avoided.

f. Names which are very long or difficult to pronounce in Latin should be avoided, and abbreviated names are preferred.

g. A name similar to or derived from the epithet of one of the species of the taxon should not be used.

h. Generic names should not be formed by combining words from different languages, or by combining parts of two existing generic names as they likely to be confused with names of intergeneric hybrids.

i. Adjectives used as noun should be avoided for naming generic names.

j. It is recommended not to dedicate generic names to person quite unrelated to Botany, and names can be suggested in honour of the great Botanists e.g., *Linnaea* after Linnaeus or atleast persons associated with natural science

k. It is to be given a feminine form to all personal generic names, whether they commemorate a man or a woman.

2. While naming after a person, the name should be formed as given below:

If the name of a person ends in	**The following letter(s) is added to form generic name:**
i. Vowels except 'a'	letter 'a' is added. E.g., Santapaua after the name Santapau which ends with a vowel 'u'.
ii. Person's name if ends in 'a'	the letters 'ea' are added. E.g., Collaea after the person's name Colla.
iii. Name ends in 'ea'	in such case no letter is added. E.g., Correa after Correa.
iv. Consonants except 'er'	the letters 'ia' are added.e.g., Adansonia after Adanson ends with consonants 'on'.
v. Name ends in 'er'	letter 'a' is added e.g., Kernera after Kerner ends with 'er'.

vi. Name ends in 'us' (in Latinized names)	if a person's name ends with 'us' the termination 'us' is removed before adding the suffix 'a' e.g., Dillenia after Dillenius

3. The type of a genus is a species and hence the type species of the new genus should be clearly indicated. For e.g., the type of the genus *Vernonia* is *Vernonia noveboracensis* (L.) Michx. If not already described and designated 'type species' should be described.

4. All procedures have to be strictly followed for valid publication of a new species.

New species At the fourteenth International Botanical Congress adopted in Berlin Code (1988), under section 4 of chapter III, it was stated that, the name of a species is a binary combination consisting of the name of the genus followed by a single specific epithet (Art.23). e.g., *Euphorbia hirta*, where *Euphorbia* denotes the generic name followed by a specific epithet *hirta*.

Once the taxon is identified as a new species, the concerned author(s) has to follow certain taxonomic and nomenclatural procedures as required by the Code.

For selecting the specific epithet for a new species the following points should be considered:

1. i. Specific epithet must consists of a single word; if it has more than two words,it must be united or should be connected by a hyphen. e.g., nux-vomica (*Strychnos nux-vomica*). As far as possible the hyphenated words should be avoided.

   ii. Specific epithets may be taken from any source whatever, and may even be composed arbitrarily.

a. Their may be commemorative names in honour of some eminent persons associated with Botany. In such case the names must be of either genitive or adjectival form. e.g., senguptae after Sengupta (*Euphorbia senguptae*) or based on any character like viridis- green colour (*Amaranthus viridis*) or black colour like nigra (*Brassica nigra*) or any locality from where it was collected e.g., indicus after India (*Tamarindus indicus*) or any habitat e.g., aquatic (*Veronica aquatica*) etc.

b. Adoption of a specific name or an epithet which has been previously used for a different species in the same genus must be avoided.

c. It is recommended under Article 23 of the Code (Berlin Code 1988) that the specific epithet should not exactly repeat the generic name like *Malus malus*, as it becomes a tautonym which is illegitimate and to be rejected.

d. A name which is very long or difficult words must not be chosen as it will be difficult to pronounce in Latin.

e. Specific epithets which exhibit a character common to all or nearly all the species of a genus may be avoided.

f. The epithets which are very much alike when used in the same genus should be omitted.

g. Specific epithets should not be formed by combining words from different languages.

h. The selected specific epithets must be with Latin terminations as far as possible.

i. The names that are little known or very restricted localities must be avoided, unless species are quite local.

j. Specific epithet named

- ★ After a place or locality is preferably an adjective and it ends with the following terminations: -ensis, -nus, -ianus, or –icus. E.g., 1. *Hevea brasiliensis*. after Brazil . 2. *Pogostemon travancoricus* after Travancore,
- ★ Names of men and women and also countries and localities may be substantive in the genitive and adjectives.
- ★ Specific epithet, when adjectival in form and not used as substantive, agrees grammatically with the generic name, e.g., *Brassica nigra*.

k. When the specific epithet is named after a man, the name should be formed by the following terminations in genitive cases are as follows:

If the name of a person ends in	**The following letter(s) is added to form specific name**
i. Vowels except 'a'	the letter 'i' is added. E.g., Euphorbia santapaui after Fr. Santapau.
ii. 'a'	the letter 'e' is added. E.g., Euphorbia senguptae after Sengupta.
iii. Consonants except 'er'	the letters 'ii' are added. E.g., Polygala jacobii after Jacob
iv. 'er'	the letter 'i' is added. E.g., Apama barberi after Barber

l. If the name of a person is already in Latin or Greek, the appropriate Latin genitive should be used. For e.g., alexandri from Alexander, augusti from Agustus, linnaei from Linnaeus.

m. While naming after a person the specific name can also be formed by adopting the terminations in the adjectival form based on the gender of the generic names in the following ways:

If the name of a person ends in	The following letter(s) is added to form specific name
i. Vowels and 'er'	an(a); an(us); an(um). E.g., 1.Jatropha heyneana after Heyne 2.Lasianthus blumianus after Blume. 3.Biophytum candolleanum after de Candolle. 4.Verbena hasslerana after Hassler.
ii. Consonants except 'er'	ian(a); ian(us); ian(um). e.g.,1.Crotalaria grahamiana after Graham. 2.Calamus huegelianus after Huegel. 3.Ageratum houstonianum after Houston.

2. Specific epithet should be followed by the original author(s) name(s). Author's name must be abbreviated, unless they are very short and followed by the date of publication.
   E.g., Theobroma cacao L.

   Pyrus malus Linn.(1753).
3. Rank of the taxon should be clearly indicated followed by the original author's name and date of publication.

4. A good detailed description for the plant has to be prepared as it will be the original description.
   * Description of the morphological characters of the group must be documented in an order.
   * The description of a plant species should not be a very lengthy at the same time a short note, which may lead to omission of certain data. Therefore, the description of a species should be fairly completed by stressing all the important characters that are essential for it.
   * The style should be telegraphic, concise with large nouns and adjectives in the nominative case.
   * While describing, grammatical sense should be maintained strictly by using comma(,), semicolon(;), colon (:) and full stop(.).
   * Description of the plant parts should follow a conventional sequence followed in taxonomic study. Therefore, it is usual to describe the nature of the habit and underground parts first. It is then followed by description of leaves, inflorescence, flowers etc. The floral parts are described in the order of calyx, corolla, androecium and gynoecium. The fruits and seed characters are described at the end.
   * Measurement of parts should be given in the metric system. The range of measurements of a part or average measurements are to be given.
5. Authors should give a full description or at least a diagnosis for the new species in Latin. The diagnosis should mention the points of interest in which the new species differs from its allies. However, it is recommended by the Code that the names published on or after 1.1.1935 must be accompanied by a Latin description or diagnosis, therefore both the diagnosis and description must be in Latin for convenience.

6. In case of parasitic plants, the description or diagnosis should always be followed by the clear indication of its nature (Root or Stem parasite, total or partial parasite) and hosts.

7. In addition, the author should supply the necessary figures with details of structure as an aid to identification and also the scale of the figures should be clearly indicated.

8. It is the recommendation of the Code that, when the names published on or after 1.1.1958 are valid only when the nomenclatural type is indicated followed by description or diagnosis and the place where the type specimen is permanently deposited and preserved should be clearly indicated (Art. 37). The indication of type specimen should be in Latin.therefore the description of a new species must be followed by the indication of type, citation of author, collection number, date, and with other particulars of its collections.

9. The new species should be published in a printed matter and must be published in a well known and widely circulated scientific journals.

REVIEW QUESTIONS

CHAPTER-1

INTRODUCTION

1. The language for biological nomenclature is ....
2. The ICBN is published by ....
3. The forerunner of Index Kewensis is....................
4. 'Theorie elementaire de la botanique' was published by ....
5. The Paris Code (1867) is otherwise called ....
6. The rules for publication of plants are given in ....
7. 'Nomenclator Botanicus' was written by ....
8. Heywood proposed Uninomial system in the year............
9. The rules for naming of cultivar is regulated by ....
10. The concept of binomial nomenclature was first introduced by...............
11. Any taxonomic group in plant taxonomy is called............
12. The first International Botanical Congress was held in ....
13. ' Nomenclator Botanicus' was published in the year ....

14. The binomial nomenclature practised even today in many countries was properly introduced by.......................
15. Linnaeus introduced binomial nomenclature in his ....
16. The rules for naming of plant species is determined by
    a. International Code of Botanical Nomenclature
    b. International Union for Plant Breeders
    c. International Code of nomenclature for cultivated plants
    d. International Plant Breeders Rights
17. Plant names are proposed in Latin language because
    a. Latin is a dead language
    b. Latin is a simple language
    c. Latin is the language of many scientists
    d. Latin is an ancient language
18. The main reason for giving scientific names to plants is
    a. to communicate universally
    b. to give difficult words to people
    c. to create confusion to students
    d. to give knowledge to tribals
19. What is botanical nomenclature?
20. Write a note on the Linnaean Principles of botanical nomenclature.
21. Why latin language is selected for botanical nomenclature?
22. Mention any two works of Linnaeus.
23. Write a note on uninomial system of nomenclature.
24. "The plants must be named with universally accepted scientific names" – Discuss.

CHAPTER 2

POLYNOMIAL SYSTEM OF MOMENCLATURE

1. 'Pinax' was written by ....
2. What are Polynomials ? Give an example.
3. Explain Polynomial system of nomenclature.

CHAPTER-3

BINOMIAL SYSTEM OF NOMENCLATURE

1. The Species Plantarum was published by ....
2. The binomial system is used for naming of ....
3. The system of naming of plants was proposed by ....
4. Linnaeus published "Binomial nomenclature" in his
    a. Origin of Species
    b. Genera Plantarum
    c. Species Plantarum
    d. Index Kewensis
5. A botanical name of a species is a combination of a generic name and a specific epithet. True/False.
6. Define binomial system of nomenclature.
7. Trigonella foenum-graecum. Explain the use of hyphen in the citation.
8. Explain binomial nomenclature and its significances.

CHAPTER-4

UNINOMIAL SYSTEM OF NOMENCLATURE

1. What is uninomial system of nomenclature?
2. Who proposed uninomial system of nomenclature?
3. What is Reference Number?
4. Explain uninomial system and its advantages.

CHAPTER-5

HISTORY AND DEVELOPMENT OF THE CODE OF BOTANICAL NOMENCLATURE

1. The botanical rules framed by Alphonse de Candolle is called ....
2. Saint Louis Code was proposed in the year..............
3. The first International Botanical Congress was headed by.............
4. The word 'autonym' was introduced in ....Code.
5. The botanical Code that published in the year 1988 is called ....
6. 'Guide for determination of types' was proposed in ....Code.
7. 'Lois de la Nomenclature Botanique' was published by.............
8. Tokyo code was published in the year ....
9. The term 'taxon' was coined by

a. Carolus Linnaeus

b. Tournefort

c. Lam

d. **Adolf Meyer**

10. The first botanical congress was held in Paris in the year

a. **1867.**

b. 1999.

c. 1988.

d. 2005.

11. The word 'taxon' was introduced in

a. Tokyo code 1994.

b. Rochester code 1982.

c. **Paris code 1956**.

d. Berlin code 1988.

12. The "type concept" was first introduced in

a. **Rochester Code**.

b. Vienna code

c. Cambridge code

d. Leningrad code

13. 'Candollean Code' is otherwise called ....

14. List out the principles of botanical nomenclature.

15. 'Philosophia Botanica' was published by ....

16. Write a note on American Code.

17. Write a critical note on 'Kew Rule'.

18. Write the importance of 1.1.1932 in plant taxonomic literature.

19. Write an essay on History of Botanical Nomenclature and its development.

CHAPTER-6

MAIN OUTLINE OF SAINT LOUIS CODE (2000)

1. Give the main outline of St. Louis Code.
2. Listout the principles of ICBN.

CHAPTER-7

RANKS AND NOMENCLATURE OF TAXA

1. The ranks of hybrid taxa are called ....
2. The family Magnoliaceae is a primitive family. True / False.
3. The name of a genus is always uninomial. True / False.
4. The family Asteraceae are considered as an advanced family. Is the sentence correct or not. Why?
5. The use of the term taxon for plants is proposed by

   a. **Lam** b. Meyer c. Tournefort d. Linnaeus
6. The name of the author(s) of the species is always printed in

   a. Bold italics type

   b. Italic type

   c. Bold type

   d. **Roman type**

7. The taxonomic term that may be substituted for any rank is called
   a. **Taxon**
   b. Class
   c. Species
   d. Order
8. Pickout the correct descending order of hierarchial ranks from the following:
   a. Family, order, species, genus.
   b. **Division, Class, Subclass, order.**
   c. Subclass, Division, Family, Species.
   d. Species, order, Family,Class.
9. The suffix used to denote family name is
   a. oidea,
   b. opsida
   c. **aceae**
   d. oineae
10. The examples of unigeneric families are
    a. Annonaceae & Cruciferae
    b. Typhaceae & Asteraceae
    c. Orchidaceae & Cannaceae
    d. **Typhaceae &Cannaceae**
11. The term species was coined by
    a. Aristotle
    b. Carolus Linnaeus
    c. Theophrastus
    d. **John Ray.**

12. A group of similar plants which breed true among themselves is called

    a. Family

    b. **Species**

    c. Order

    d. Genus

13. A group of genera which show resemblances to one another in their morphology is termed as

    a. Class b. Order c. Tribe d. Family

14. Choose the correct descending sequence of taxonomic categories from the following:

    a. class, division, genus, order, tribe, family

    b. family , tribe, genus, order, division, class

    c. class, order, tribe, family, genus, division,

    d. **division, class, order, family, tribe, genus.**

15. Pickout the suffix that characterizes the rank of an order.

    a. aceae b. ales c. oideae d. neae

16. Any taxonomic categorie ending with 'aceae' indicates the name of

    a. Family b. Subtribe c. genus d. Order.

17. Select the ascending order of taxonomic ranks from the following:

    a. **Species, Genus, Family, Order**

    b. division, Group, species, family

    c. family, order, species, division

    d. species, order, subclass, family

18. Define the term taxon.

19. Define family.

20. What is rank ?
21. Define species.
22. What is hierarchial system of classification?
23. Write the taxonomic categories for *Rosa indica*.
24. Explain the structure of taxonomic hierarchy.
25. Write the list of conserved family names and their revised names.
26. Discuss in detail the rules of ranks and nomenclature of taxa described in the code.

CHAPTER-8

THE TYPE METHOD

1. The type of a family is a.............
2. Duplicates of holotype is called ....
3. The type of a genus is a ....
4. Application of names of taxa on the basis of nomenclatural type was given by ....
5. The nomenclatural types are introduced by ....
6. What is typification?
7. What is type locality?
8. What is type concept?
9. What is nomenclatural type?
10. What is the significance of a type specimen?
11. What is holotype? Give an example.
12. What is an isotype? Give an example.

13. What is paratype? Give an example.
14. What is syntype? Give an example.
15. What is lectotype? Give an example.
16. What is neotype? Give an example.
17. How will you differentiate holotype from neotype?
18. How will you differentiate neotype from lectotype?
19. Distinguish between syntype and holotype.
20. What is an autotype ? Cite an example.
21. How will you select lectotype ?
22. What are the conditions needed for selecting neotype ?
23. Explain the determination of different types with suitable examples.
24. Explain how the type method works in solving taxonomic problems.

CHAPTER-9

PRINCIPLES OF PRIORITY AND THEIR LIMITATIONS

1. The principles III and IV of the ICBN deal with principles of priority. True/ False.
2. The rule of principle of priority cannot be applied to ....
3. Name the first species conserved by Berlin code (1988).
4. Sterculiaceae Lindl. (1830) nom. cons. - Explain.
5. Listout the limitations of priority.
6. How will you cite the names of conserved species?

7. Explain principles of priority and their limitations.
8. Write the botanical name of coral tree.
9. What is principle of priority ?
10. "Principle of priority is important in botanical nomenclature". Discuss.
11. Give an example of tautonym.
12. Write the binomial of apple.
13. Write the importance of 1st May 1753 in taxonomic literature.

CHAPTER-10

EFFECTIVE AND VALID PUBLICATION

1. A plant name published without description is termed as ....
2. Publication of a name is effected only by distribution of printed matter. Is it correct ?
3. The rejected names are
   a. legitimate names
   b. conserved names.
   c. valid names.
   d. illegitimate names.
4. What is effective publication?
5. What is valid publication?
6. Write the importance of 1st Jan. 1953 with reference to publication of plant names.

7. Write the importance of the following dates in taxonomic literature.
   i. 1.1.1953
   ii. 1.1.1973
   iii. 1.1.1996
   iv. 1.1.1958
8. Explain the rules of effective and valid publications of species.

CHAPTER-11

CITATION OF AUTHORS

1. What is citation of author ?
2. The binomial of mango is *Mangifera indica* L. The letter 'L' in the citation stands for ....
3. *Lupinus* [Tourne] Linn. Explain the use of square bracket in the above citation.
4. Write the significance of author's citation.
5. Explain the following :
   a. *Indotristicha tirunelveliana* Sharma *et al.*
   b. *Pennisetum typhoides* Stapf et Hubbard.
   c. *Cassia montana* Heyne ex Roth.
   d. *Hopea wightiana* Wight & Arn.
6. Write the use of the following words in citation of names with suitable example.
   i. ex.
   ii. et al

iii. &

iv. in

v. emend

7. Explain the following words in the citation of plant names with examples.

   a. nom .cons.

   b. nom.nud.

   c. auct non.

   d. parenthesis

8. Explain the rules of publication of later homonym with an example.

9. Explain the rules of valid publication of species.

CHAPTER-12

RETENTION OF NAMES AND EPITHETS OF TAXA

1. The articles dealing with the rules of retention of plant names are

   a. 60-67

   b. 51-56

   c. 65-70

   d. 25-30.

2. *Myosotis* L. emend. R. Brown. Explain the use of the word 'emend' in the citation.

3. Explain the conditions required for the retention of original specific epithet, when the species is transferred to another genus.

4. Explain in detail the rules for retention of various plant names with suitable examples.
5. Explain the following:
   a. *Portulaca pilosa* subsp. *pilosa*
   b. *Gomphandra* Wall. ex Lindl. emend Becc.
   c. *Gypsophila* sect. *vaccaria* (DC.) Godr.
   d. *Centella asiatica* (L.) Urban.

CHAPTER-13

CHOICE OF NAMES

1. Mention the article numbers that deal with the rules for choosing the correct names.
2. When the genus *Derris* Lour. (1790) and *Pongamia* (1803) are merged, what will be the correct name ?
3. Explain the conditions needed for selecting the correct names.
4. When the species *Waltheria indica* L. (1753) was united with *Waltheria americana* L. (1753), Robert selected the name *Waltheria indica* L. as the correct name for the combined species. Was his choice correct ? Explain.
5. A non- fossil genus *Platycarya* Sieb & Zucc. (1843) was united with a fossil genus *Petrophiloides* Bowerbank (1840) . Select the correct name giving reasons.

CHAPTER-14

REJECTION OF NAMES AND EPITHETS

1. What is legitimate name ? Give an example.
2. What is illegitimate name? Cite an example.
3. Give an example of a naked name.
4. What is monstrosity? Give an example.
5. What are naked names? Cite an example.
6. Define superfluous name.
7. What are tautonyms? Cite an example.
8. What are homonyms? Give example.
9. Write the conditions needed for the use of the words 'nom.nud'. in the citation of names.
10. Differentiate nomenclatural and taxonomic synonyms.
11. Distinguish between synonyms and homonyms with examples.
12. Define the following terms :
    a. Basionym
    b. Ambiguous name
    c. Doubtful name
    d. Autonym
    e. Nomina rejicienda
13. d. Homonym
14. Write a critical note on synonyms.
15. Write a short note on illegitimate names.
16. Give an account on various illegitimate names mentioned in the code with examples.

17. *Astragalus rhizanthus* Boiss. (1843) is considered as a later homonym of *Astragalus rhizanthus* Royle (1835). Is it correct ? Why?

CHAPTER-15

ORTHOGRAPHY OF NAMES AND EPITHETS

1. Select the correct name from the following:
   i. *Amaranthus* L. 1753. and
   ii. *Amarantus*.
2. Explain orthographic names of error with suitable example.
3. Write a note on orthographic variants.

CHAPTER-16

NOMENCLATURE TERMINOLOGY

1. Define valid name.
2. What are alternative names? Give suitable examples.
3. What is a correct name?
4. What are rejected names? Give an example.
5. What are naked names? Cite an example.
6. Tautonyms are rejected by the code. True/False
7. What are confused names?
8. What is 'nomen novum'?

9. What are ambiguous names?
10. What is Latin diagnosis?
11. Explain parenthetical authority with suitable example.
12. *Brassica rapa* L. subsp. *rapa*. – Explain.
13. Critically comment on synonym.
14. Write a critical note on autonym.
15. Write the alternative name of the families Labiatae and Cruciferae.
16. What are tautonyms ? Give an example.
17. Why are names conserved by the Code ?
18. What is epithet briging name ?

CHAPTER-17

NOMENCLATURE OF HYBRIDS AND CULTIVATED PLANTS

1. ICNCP govern the rules for hybrids and cultivated plants. True/False.
2. What is a hybrid ?
3. Expand ICNCP.
4. Hybridity is indicated by the use of
    a. **'×' sign**
    b. '–' sign
    c. '÷'sign
    d. '+'sign
4. What is hybrid formula ? Cite an example.

5. What is a cultivar? Give an example.
6. Write a critical note on graft hybrids.
7. What is nothotaxa? Give an example.
8. What is 'collective epithet' ?
9. What are nothospecies?
10. How will you cite the graft hybrids ?
11. What is condensed formula? Give an example.
12. Write the abbreviation used for cultivars.
13. Distinguish between interspecific and intergeneric hybrids.
14. Write a note on cultivated plants.
15. Explain the rules and regulations mentioned by the code to name hybrids and cultivated plants.
16. Explain the following:
    - *Musa* × *paradisiaca*
    - *Digitalis purpurea* × *D. lutea*
    - *Salix* × *capreola* (Salix aurita × S. caprea)
    - × *Triticale* (*Triticale* × *Secale*)
    - × *Vascostylis* Takakura (=*Ascocentrum* × *Rhynchostylis* × *Vanda*)
    - *Rosa webbiana* + *R. floribunda*
    - *Citrullus lanatus* cv. Sugar baby
    - *Rosa floribunda* 'Blessings'

CHAPTER-18

NOMINA CONSERVANDA

1. What is nomina conservandum ?
2. Why are names of families and genera conserved by the code?
3. The concept of nomina conservanda was introduced at the third IBC at ....
4. Write an account on nomina conservanda.
5. Write the guiding principles of conservation of names.
6. Write a short note on nomina generica conservanda.
7. Comment on nomina specifica conservanda.
8. Write a note on nomina familiarum conservanda.

CHAPTER-19

PUBLICATION OF A NEW SPECIES

1. Outline the procedure for publication of a taxon.
2. Write an essay on the rules of publication of a new species.

EXERCISE

1. *Thea* L. (Sp.Pl. 1:515, May 1, 1753) and *Camellia* L. (Sp. Pl. 2: 698, Aug. 1, 1753), are considered congeneric. Is it possible to decide the correct name of the genus.

   Ans. Yes. The generic names *Thea* L. and *Camellia* L. are treated as having been published simultaneously in May 1753. The above example indicates that, the code does not give importance to priority of publication, when both names of taxa are published on the same date (1753).

   Hence, the author who first unites these two taxa can free to choose any one of them as the correct name and his choice must be strictly followed. Accoringly when the author Sweet, first united these two genera, selected *Camellia* L. (1753) as the correct name and cited *Thea* L. as a synonym and his choice should be followed.

2. Explain the following:

   a. *Eclipta prostrata* (L.) L. (1771).

   Ans. In the above citation, Linnaeus would have given earlier a different name to this plant. Later on in 1771, he himself would have transferred that species to the genus *Eclipta* and retained the same specific epithet under the genus *Eclipta*.

   According to the rule of nomenclature the author (Linnaeus) of the basionym (epithet bringing name) must be written within bracket, and outside bracket the author (Linnaeus) who transferred the name must be cited as written below:

Eclipta prostrata (L.) L. (1771).

Note: Publication of the original name as well as the transfer of the name was made by the same author (Linnaeus).

b. *Rosa gallica* L. var. *gallica*

Rosa gallica var. *eriostyla* Keller

Ans. The above citations indicate that the species *Rosa gallica* was further reduced to two varieties by Keller. According to the Art. 52.1 of the code, when a species is divided into two or more varieties, there should be a recognition of a variety of the species *Rosa gallica* that does not contain its type, automatically establishes a variety that does contain the type. Hence, the variety name is written without the citation of author's name at the end of citation as given in the first citation. (since the name *gallica* was already published by Linnaeus, therefore the author Linn. need not be repeated again). Whereas in the second citation, a new variety eriostyla was proposed by Keller. Therefore, the author's name is cited after the name of the variety *eriostyla* because Keller found this *eriostyla* as a new variety, but not the type of variety of gallica of Linnaeus. Hence the new variety of *Rosa gallica* must be written with authors' citation as given in the second citation. Eg., *Rosa gallica* var.*eriostyla* Keller.

c. *Saponaria* sect. *vaccaria* DC. when transferred to *Gypsophila* becomes *Gypsophila* sect. *vaccaria* (DC.) Godfr.

Ans. *Saponaria* sect. *vaccaria* was first described and published by DC. Later on Godfr, who transferred *Saponaria* sect. *vaccaria* to the genus *Gypsophila*. While transferring the genus *Saponaria* to *Gypsophila* the name of the specific epithet or sectional name if valid, it must be retained and the also the author of the basionym (DC) must be written within parenthesis and it is then followed by the name of the author who made the transfer.

Hence the citation is written as below:

Gypsophila sect. *vaccaria* (DC.) Godfr.

It is therefore example of double citation, parenthetical authority, and also for generic transfer.

d. *Antirrhinum spurium* L. (1753) when transferred to the genus *Linaria* was correctly called *Linaria spuria* (L.) Mill. (1768), while *Spergula stricta* Sw. (1**799**) when transferred to the genus *Arenaria* could not be correctly called *Arenaria stricta* (Sw.).

Ans. *Antirrhinum spurium* L. (1753) when transferred to the genus *Linaria*, was correctly called *Linaria spuria* (L.) Mill. (1768), this is because, the specific epithet *spuria* was not preoccupied in the genus *Linaria*. It is an example of generic transfer, that is from one generic level to another genus. While transferring, the name of the specific epithet, if valid (if it is not pre-occupied or become a tautonym) must be retained and the author of the basionym (Linnaeus who published *A. spurium*) should be written within parentheses and is followed by the author (Mill.), who made the transfer, hence it is written as *Linaria spuria* (L.) Mill. (1768).

While *Spergula stricta* Sw. (1**799**) when transferred to the genus *Arenaria* must be called *A. uliginosa Schleich ex Schlechtend* (1808) and not *Arenaria stricta*, as the specific epithet stricta must have been preoccupied in the genus *Arenaria*.

e. *Phyllanthus* L. emend. Muell. – Arg. --

Ans. *Phyllanthus* was originally described by Linnaeus. As the original description of this taxon published by Linnaeus was incomplete, and subsequently corrected or the circumscription of the taxon *Phyllanthus* might have been altered retaining the same nomenclatural type by Muell. - Arg--. Hence the citation must be written with the name of the original author (L.), but followed by a word 'emend' and then the name of the authors (Muell.-Arg--.) who were responsible for the revision of the work.This is an important element of revisionary work.

f. *Ficus stortophylla* Warb. (1904);

Ficus irumuensis De Wild. (1922).

Ficus exasperata auct. non. Vahl; De Wild. (1899).

Ans. In the above citations, *Ficus stortophylla* Warb. (1904), is the correct name, and *Ficus irumuensis* De Wild. (1922) is the synonym. Whereas *Ficus exasperata* auct. non. Vahl; De Wild. (1899) is an example of misidentified name. This is because, in the citation a misapplied name must be indicated by the words auct. non. followed by the name of the original author.

g. *Lindera* Thumb. (1783), non. Adans. (1763).

Ans. Earlier homonyms may be distinguished by the citation of their authorities and dates, and if necessary, by the use of the conjunction non.

In the above example, the taxon described under the name *Lindera* by Thumb. in 1783 was considered as a correct name but not the name which was proposed by Adans. in 1763 (earlier homonym) and it should be rejected.

h. *Uvaria heteroclita* Roxb. {Hort. Beng. (1813), nom. nud.; Fl. Ind. 1832}.

Ans. Names which are not accompanied by description called nomina nuda (nom. nud.) or naked names or names without description. The above cited plant name *Uvaria heteroclita* was published in Flora Indica in 1832 by Roxburgh (1813) in his work Hortus Bengalensis. It was published. It is an example of nomen nudum, because the taxon was published without description and it must be rejected.

i. *Abutilon graveolens* W&A (1834) var. *hirtum* (G.Don) Masters(1874).

A. hirtum G.Don (1831)

Ans. *Abutilon hirtum* was described as type by G.Don in 1831. But later in 1874, Masters considered the species *A. hirtum* of G.Don as a variety under the species of *Abutilon graveolens* of W&A (1834). Hence the citation is written as

Abutilon graveolens W&A (1834) var. *hirtum* (G.Don) Masters(1874).

j. *Emblica officinalis* Gaertn. (Fruct. 2:122. 1791) is a new name for *Phyllanthus emblica* L. (SP. Pl. 982. 1753).

Ans. *Phyllanthus emblica* L. when transferred to the genus *Emblica*, must be called *Emblica officinalis* Gaertn. (1791), because Gaertner could not retain the specific epithet '*emblica*', as the resulting binary name *Emblica emblica* is a tautonym which is inadmissible in accordance with the code. Gaertner therefore called the plant by the new name (nom. nov.) *Emblica officinalis* Gaertn.

3. Select the correct name from the following:

Herba turcica (Boiss.) Bornm. (1940)

Planta turcica Boiss. (1867)

Herba turcica var. *turcica* (Boiss.) Hall. (1925)

Herba neglecta Brown (1959).

Ans. *Herba turcica* (Boiss.) Bornm. (1940) is a correct name.

Planta turcica Boiss. (1867) is the basionym or epithet bringing synonym of *Herba turcica*, whereas *Herba turcica* var. *turcica* (Boiss.) Hall. (1925) is an autonym and *Herba neglecta* Brown (1959) is the synonym.

4. Write down the resulting name combinations and author citations giving reasons in the following cases :

    a. The name *Gossypium tomentosum* had been proposed by Nuttal, but not published; later it was published by Seemann (1865).

Ans. When an author who first validly publishes a name (Seemann.) ascribes it to another person who proposed the name, but not published, the correct author citation is the name of the author who proposed the name (Nuttal) followed by the word ex and then the name of the author who validly published it. Accordingly the binomial of the plant is cited as

Gossypium tomentosum Nutt. ex Seem.

or the plant name may be directly followed by the name of the author who validly published as

Gossypium tomentosum Seem.

In the citation given above, the author Seemann was responsible for valid publication of *Gossypium tomentosum*, which was first originally coined by Nuttal.

    b. *Lycopodium pallescens* C.Presl. was transferred to *Selaginella* by Spring.

Ans. When the species *Lycopodium pallescens* C.Presl. was transferred to the genus *Selaginella* by Spring., it becomes *Selaginella pallescens* (C.Presl.) Spring., which is an example of double citation and also forms a generic transfer. When a genus is transferred to another genus the specific name if properly published must be retained in the new combination and the author of the epithet bringing name (basionym) must be written within parentheses and it is followed by the name of the author who effected the transfer.

c. Allioni raised *Medicago polymorpha* var. *orbicularis* L. to species level.

Ans. When *Medicago polymorpha* var. *orbicularis* L. was raised to species level by Allioni, it becomes

Medicago orbicularis (L.) Allioni.

It is an example of double citation and also change in taxonomic position, i.e., the variety rank is raised to specific level.

5. Explain the difference between the following citations:

Mastixia Blume (1826) subg. *Mastixia*

Mastixia Blume (1826) subg. *Manglesia* Matthew (1976)

Ans. The genus *Mastixia* was originally described and published by Blume in 1826. Later in 1976 Matthew divided this genus into two subgenera as subgenus *Mastixia* and subg. *Manglesia*.

According to Art.52, when a genus is divided into two or more genera, the original name (*Mastixia*) is to be retained for the one, which has the type (type specimen) that is subg. Mastixia. Hence it is written without the citation of author's name as

Mastixia Blume (1826) subg. *Mastixia*

In the next citation, a new name is given by the author Matthew it self. Hence it was written with the author's citation. Therefore the citation is cited as

Mastixia Blume (1826) subg. *Manglesia* Matthew (1976)

6. Findout the valid name for the species of *Cyperus* by explaining the relevant rules of the code and also indicate the meaning of such citations.

Cyperus iria L. (Sp. Pl. 67, 1753) nom.dub.

Cyperus songaricus Karel (Bull. Soc. Mosc. 859, 1841)

Cyperus seminudus Moritzi, (Verz. Zoll. Pfl. 96, 1838) nom.conf.

Cyperus microiria Steud., (Zoll. Verz.Ind. Archip. 2:62,1800) nom.nud.

Cyperus diaphiniria Steud., (Syn.Pl.Gl.2:7, 1841) nom. illegit.

Cyperus microlepis Baker (Fl.Maurit.410,1860) non. Boeck(1857)

Ans. *Cyperus songaricus* Karel (1841) is the correct name as it is validly published in accordance with the provisions of the code.

The other names are not valid due to the following reasons and must be rejected by the code:

Cyperus iria L. (1753) is the doubtful name as the citation is followed by the words nom. dub.

Cyperus seminudus Moritzi, (1838) nom.conf. is the confused name.

Cyperus microiria Steud. (1800) nom.nud. is the name published without description.

Cyperus diaphiniria Steud., (1841) nom.illegit. is the illegitimate name which is not valid.

Cyperus microlepis Baker (1860) non. Boeck (1857) is the later homonym which is also not accepted according to the rule of ICBN.

7. Rottboell described in 1797, a species and named it under *Cyperus serotina*. Later C.B.Clarke (1893) transferred the species to the genus *Juncellus*.

   Indicate the correct citation and also the basionym.

   Ans. *Juncellus serotina* (Rott) C.B.Clarke (1893) is the correct name.

   Cyperus serotina Rott (1797) is the basionym.

8. When the genera *Kyllinga* Rottb (1797), *Mariscus* Clarke (1893), *Juncellus* Clarke (1893), *Cyperus* Linn.(1753),

and *Pycreus* Turn (1750), were all united, the resulting genus was designated as *Pycreus* Turn (1750). Is it correct? Explain.

Ans. No. The correct name for the resulting genus is *Cyperus* Linn.(1753), and not *Pycreus* Turn (1750).

According to the rule of principle of priority, on or after 1.1.1753 when two or more names apply to the same taxon, in general it is by the oldest validly published name that it should properly be known. Hence the correct name for the resulting genus is *Cyperus* Linn. (1753), and not *Pycreus* Turn (1750), as given above in the problem, which was published before 1.1.1753 has no status in the code, hence, though the name *Pycreus* was published earlier than all other names, it becomes an illegitimate name.

9. When a new subspecies is to be proposed under *Carex muricata* L. (Sp.Pl. 1753), how will you give the citation?

Ans. The new subspecies proposed under *Carex muricata* is, *Carex muricata* L. subsp. *muricata*, which is an example of an autonym. According to Art. 46 of the code, a recognition of a new subspecies that does not contain the type of the species automatically establishes a subspecies that does contain the type. In this binomial, the species muricata was already published by the author Linnaeus.

Hence the final epithet of an autonym at the infraspecific level is that of the specific name unaltered, but without the citation of an author's name.

10. *Theka* Adans. Fam. 2: 445. 1763.

Tectona L.f. Suppl. 151. 1781. nom. cons.

Both these genera are synonymous. Select the correct name giving reasons.

Ans. The correct name is *Tectona* L.f. (1781), as it is conserved against its earlier published name *Theka* Adans. (1763), due to its very long usage in botanical literature.The names are conserved by the code in order to avoid disadvantageous changes in the nomenclature of genera by the strict application of the principle of priority and also to stabilize the rules of nomenclature. According to Art. 11 of the code, for any taxon from family to genus inclusive the correct name is the earliest legitimate one with the same rank except in cases of limitation of priority by conservation. Therefore in this case the rule of priority can not be applied and the generic name *Theka* Adans. (1763), though published earlier is to be rejected.

11. In the light of ICBN, explain the steps that you could adopt under the following conditions:

    i. When *Chloris radiata* (L.) Sw. (Prodr. 26, 1788) based on *Agrostis radiata* L. (Syst. Nat. Ed. 10, 2:873, 1959) is considered to be nomenclaturally superfluous.

    Ans. When a name published for a taxon, which includes the type material of another taxon of the same rank, for which there is already a legitimate name available is called 'nomenclaturally superfluous' and it should be rejected. Hence the name *Agrostis radiata* L. 1959 is considered to be nomenclaturally superfluous and must be rejected.

    ii. Are the following binomials tautonyms?

    Malus malus (L.) Britton

    Linaria linaria

    Bambusa bambos and

    Cajanus cajan

Ans. A binary name in which the specific epithet exactly repeats the generic name is called a tautonym. It is treated as an illegimate name by the code and must be rejected. In the above citations the binomials such as *Malus malus* (L.) Britton and *Linaria linaria* are considered to be tautonyms and must be rejected. Whereas the other two names *Bambusa bambos* and *Cajanus cajan* are not tautonyms, as the specific names of these two species do not exactly repeat the generic names.

12. Write the bibliographic citation for the following:

 a. Saxifragaceae in Fl. Brit. India. Hooker. J.D. 1855.

 Ans. Hooker, J.D. Saxifragaceae in Fl. Brit. India. London. 1855.

 b. Hutchinson, Families of Flowering Plants. (1973). Oxford, 3rd edn. Oxford University Press.

 Ans. Hutchinson, J. (1973). The Families of Flowering Plants.

 3rd edn.Oxford University Press, Oxford.

13. Select the correct name and the basionym from the following citations.

 Pilea melastomoides (Poir.) Wedd. [Ann. Sci Nat. (Ser. 4) 1:186. (1854)].

 Urtica melastomoides Poir. (Lamk. Encyclo. Suppl. 4: 223. 1816).

 Pilea trinervia Wight, (Icon.1853).

 Ans. *Urtica melastomoides* was first validly published by Poirret in 1816. In 1854 when, Weddell transferred the genus *Urtica* to *Pilea*, the specific epithet '*melastomoides*' must be retained if valid and the author of the basionym (Poirret) must be cited in Parantheses, followed by the name of the author who effected the transfer. (Weddell).

Hence it becomes *Pilea melastomoides* (Poir.) Wedd. (1854), which is a correct name.

Urtica melastomoides Poir. 1816. therefore becomes a basionym.

Whereas *Pilea trinervia* Wight (1853) forms the taxonomic synonym of *Pilea melastomoides* (Poir.) Wedd. as it is based on a different type.Therefore *Pilea trinervia* Wight (1853) became superfluous, as this plant already had an earlier published name *Urtica melastomoides* Poir. 1816.

14. Write the taxonomic endings for the following categories with suitable example:

    Divisio, Subdivisio, Classis, Subclassis, Subordo, Subfamilis and Tribus.

Ans.

Divisio	-	Magnoliophyta	-	phyta
Subdivisio	-	Magnoliophytina	-	phytina
Classis	-	Magnoliopsida	-	opsida
Subclassis	-	Magnolidae	-	idae
Subordo	-	Magnolineae	-	ineae
Subfamilis	-	Magnolioideae	-	oideae
Tribus	-	Magnolieae	-	eae

15. *Trigonella foenum - graecum* . Is it a binary combination?

    Ans. Yes, it is a binary combination. According to Art.33 of the code, the name of a species is a binary combination consisting of the name of the genus followed by a single specific epithet. If an epithet consists of two or more words, must be united or hyphened. Accordingly the words *foenum – graecum* are united. Hence *Trigonella foenum- graecum* is a binary combination.

16. *Elatostema lineolatum* Wight var. *lineolatum*. Explain.

    (Note the final varietal epithet repeats exactly like the specific epithet).

Ans. *Elatostema lineolatum* Wight var. *lineolatum* is an example of an autonym. That is a name automatically established for a taxon, merely as a consequence of the proposal of another taxon. The recognition of a variety of *Elatostema lineolatum* that does not contain its type, automatically establishes a variety that does contain the type. Hence while indicating the typical variety that contains the nomenclatural type of the species, the specific epithet '*lineolatum*' has been repeated unaltered but without writing the author's name (Wight) as cited below:

Elatostema lineolatum Wight var. *lineolatum*.

17. Correct the following sentences wherever necessary:
    i. Arecaceae is an advanced family.
    ii. The subfamily papilionoideae has many genera.
    iii. The order Rosales differs from parietales.
    iv. The genus name *Citrus* is considered as singular.
    v. The family Euphorbiaceae is distinguished by unisexual flowers.
    vi. The family Asteraceae consists of large number of genera and species.

Ans. i. Arecaceae **are** advanced family.

ii. The subfamily papilionoideae **have** many genera.

iii. The order Rosales **differ** from parietales.

iv. The genus name *Citrus* **is** considered as plural.

v. The family Euphorbiceae **are** distinguished by unisexual flowers.

vi. The family Asteraceae **consist** of large number of genera and species.

18. The family name Ebenaceae Giirke, nom. cons. is based on the stem of the illegitimate generic name *Ebenus* O.Ktze., non L;

    State whether the name Ebenaceae is legitimate or not?

    Ans. The family name Ebenaceae is legitimate because it is conserved. According to Art.18.3, a name of a family based on the stem (type) of an illegitimate generic name is illegitimate unless conserved.

19. State whether the procedure given below is correct or not ? Explain.

    a. The genus *Priva* was divided into three genera *Blairia*, *Burseria* and *Priva*.

    Ans. Yes. The above procedure given is in accordance with the rules of the botanical nomenclature. According to Art. 52.1, when a genus is divided into two or more genera, the generic name which has the type specimen must be retained for one of them. Hence, the genus name *Priva* is retained for one of them, which includes its type.

    b. The genus *Euphorbia* L. is divided into genera like *Chamaesyce* Gray, *Poinsettia* Grah. and *Tithymalus* Gaertn.

    Ans. The above procedure followed is contrary to the rule of nomenclature. According to the code, when a genus is divided into two or more genera, the generic name which has the type must be retained for one of them. But the rule was not followed in the above procedure in dividing the genus *Euphorbia*. Hence, the name *Euphorbia* L. must be kept for one of the genera which includes the type of *Euphorbia* L.

20. When the three genera – *Sloanea* L. (1753), *Echinocarpus* Blume (1825) and *Phoenicosperma* Miq. (1865) are united, which name should be adopted for the resulting genus? Explain.

Ans. According to the rule of priority when two or more names based on the same type bearing different dates are united, it is the oldest name published on or after 1.1.1753 should be selected as the correct name for the combined genus.Therefore correct name is *Sloanea* L. (1753), as it is the oldest name of these three generic names and hence it should be adopted for the resulting genus.

21. *Dregea angustifolia* (Hook.f) Santapau & Irani (Bot. Mem. Univ. Bombay 4: 41. 1960.

    D. volubilis var. *angustifolia* Hook.f. (Fl.Brit. India 4:47. 1883.)

    Marsdenia lanceolata Cooke, (Fl. Pres. Bombay 2:167). 1904.

    Dregea lanceolata (Cooke) Santapau & Wagh. (Bull. Bot. Surv. India 5: 106) 1964.

    What is the correct name of the above taxon when treated as a species of *Dregea* ?

    Ans. *Dregea lanceolata* (Cooke) Santapau & Wagh. (1964) is the correct name of the taxon when it is treated as a species of *Dregea*. According to art. 60 of the code, when the rank of a genus or infrageneric taxon is changed, the correct name is the earliest legitimate one available in the new rank. Accordingly when Santapau & Wagh. transferred *Marsdenia lanceolata* to the genus *Dregea* retained the specific epithet '*lanceolata*' in *Dregea*. Therefore the species *Dregea lanceolata* (Cooke) Santapau & Wagh. (1964) is the correct name.

    In no case does a name or an epithet will have priority outside its own rank. Whereas the species *Dregea angustifolia* (Hook.f) Santapau & Irani 1960 though published earlier could not be considered as a correct name because, the varietal epithet '*angustifolia*' has no priority outside its own rank and hence the combination

Dregea angustifolia (Hook.f) Santapau & Irani (1960) is illegitimate and must be rejected.

22. What is the alternate name for Palmae and Guttiferae.

    Ans. The alternate name for Palmae is Arecaceae, and for Guttiferae is Cluciaceae.

23. Are the following family names correct ? Explain.

    Brassicaceae, Solanaceae, Labiatae, Aroideae, Rutaceae and Compositae.

    Ans. According to the rule of botanical nomenclature, the name of a family is formed by adding the termination 'aceae' to the stem of a legitimate name of a type genus. Therefore the family names Brassicaceae, Solanaceae and Rutaceae are correct, as the family name Brassicaceae is based on the type genus *Brassica*, Solanaceae from the stem of a type genus *Solanum* and Rutaceae from the type genus *Ruta*.

    Whereas the family names Labiatae, Aroideae and Compositae though not ending with the suffix 'aceae' and also not formed from the stem of generic names, they are also considered as correct name, because they are conserved due to their long usage in the taxonomic literature. But the code has however, permitted, the usage of alterntive names for these familes, accordingly they are called Lamiaceae from the type genus *Lamium* for Labiatae, *Arum* for Aroideae, and Asteraceae from the type genus *Aster* for Compositae.

24. *Scilla peruviana* L. (1753) is a legitimate name, but not found in Peru. Can we accept it or not? Why?

    Ans. As *Scilla peruviana* is validly published and also a legitimate name it can be very well accepted, whether it is present in Peru or not.

25. Bentham (1876) first united the genera *Ophioxylon* L. (1753) and *Rauwolfia* L. (1753) adopting the name

Rauwolfia for the combined genus. Was his selection correct ? Explain.

Ans. Yes. According to Art. 57.2, the author who first unites taxa bearing the names of the same date (published year, 1753) must choose any one of them, as a correct name and his choice must be followed. Hence, Bentham's choice of selecting the name *Rauwolfia* for the combined genus is correct and his choice should be retained.

26. *Pyrus malus* L. 1753.

Malus pumila Mill. 1768.

Malus malus (L.) Britton. 1913.

All the three names are nomenclatural synonyms, i.e. based on the same type. What is the correct name if the species is treated under the genus *Malus*? Explain.

Ans. In the above citations the species *Pyrus malus* was first published by Linnaeus in the year 1753. Miller (1768) while transferring this species to the genus *Malus* could not retain the specific epithet 'malus' as the resulting binary name '*Malus malus*' became a tautonym, which was not accepted by the code; Hence he correctly proposed a new specific epithet name called *Malus pumila* Mill. (1768). The subsequent publication of the tautonym *Malus malus* (L.). Britton (1913) is considered as illegitimate. Hence the correct name of the species when treated under the genus *Malus* is *M.pumilla* Mill.

27. *Cyperus iria* var. *rectangularis* Kukenth. (Engler, Pfreich.101:152. 1935).

Cyperus rectangularis (Kukenth.) Bennet (Indian For. 95: 692.1969).

Cyperus alulatus Kern. (Reinwardia 1:463. 1952).

What is the correct name of the taxon when treated as a species?

Ans. *Cyperus alulatus* Kern. is the correct name of the taxon when it is treated as a species. In no case does a name or an epithet will have priority outside its own rank. Therefore the varietal epithet *'rectangularis'* though published earlier, has no priority outside its own rank and hence the combination *Cyperus rectangularis* (Kukenth.) Bennet (1969) is illegitimate and must be rejected.

28. *Rauwolfia serpentina* (L.) Benth. ex Kurz, (Fl. Burma 2:171. 1877.

Basionym : *Ophioxylon serpentinum* L., Sp. Pl. 1043. 1753.

Explain the usage of L. in Parentheses and *ex* in the author citation i.e.(L.) Benth. ex Kurz.

Ans. *Rauwolfia serpentina* (L.) Benth. ex Kurz was a combination based on the basionym (epithet- bringing synonym) *Ophioxylon serpentinum* L.(Art. 33.2). When a species is transferred to another genus but retains its specific epithet if valid, must be retained and the author of the basionym must be cited in parentheses, followed by the name of the author who effected the combination (Art.49). Kurz in 1877 validly published the combination by directly giving reference to the basionym, but ascribed it to Bentham. According to Recommendation 46C I, the correct author citation is the name of the publishing author (Kurz), but the name of the other person followed by the connecting word *ex* may be inserted before the name of the publishing author, if desired, i.e., *Rauwolfia serpentina* (L.) Benth. ex Kurz.

29. *Phyllanthus* Linn. emend Mull - Arg. Explain the usage of the word 'emend' in the citation.

Ans. When the original description of a taxon is incomplete and subsequently corrected the diagnosis, or the circumscription of a taxon is altered retaining the same type, the citation of the original author is retained, but followed by the word 'emend' and also the name of the author responsible for the correction (Art.47). This is an important element of revisionary work of taxonomy.

30. Write the name of the ranks with appropriate suffix to the following taxon '*Rosa*' from Division to subtribe.

Ans.

Division	-	phyta	Magnoliophyta
Subdivision	-	phytina	Pterophytina
Class	-	opsida	Magnoliopsida
Subclass	-	opsidae	Pteropsidae
	-	idae (seed plants)	Rosidae
Order	-	ales	Rosales
Suborder	-	ineae	Rosineae
Family	-	aceae	Rosaceae
Subfamily	-	oideae	Rosoideae
Tribe	-	eae	Roseae
Subtribe	-	inae	Rosinae
Genus	-------		Rosa

GLOSSARY

Affinis(aff.) Allied to, related to.

Alternative family name (nom. alternativum) One of eight family names, each regularly formed from a generic name in accordance with Code, allowed as an alternative to one of the family names of long usage treated as validly published.

Alternative names Two or more different names based on the same type proposed simultaneously for the same taxon by the same author.

Analysis A figure or group of figures, commonly separate from the main illustration of the organism (though usually on the same page or plate), showing details aiding identification, with or without a separate caption.

Anamorph A mitotic asexual morph in pleomorphic fungi.

Apud (ap.) In the work.

Arbitrary Names formed without to etymology.

Article A series of numbered rules in the Code.

Ascription The direct association of the name of a person or persons with a new name or description or diagnosis of a taxon.

Auctorum non (auct. non.) Of others, not of.

Author citation A statement of the name(s) of the author(s) responsible for the establishment or introduction of a name; when used, it is appended to that name.

Autograph A text written by hand.

Automatic typification Typification of the name of a taxon above the rank of genus by the type of the generic name on which it is based.

Autonym An automatically established name in which a generic name or specific epithet is repeated as the final epithet in the name of a subdivision of a genus or of an infraspecific taxon that includes the type of the adopted, legitimate name of the genus or species, respectively; the final epithet of an autonym is not followed by an author citation.

[Autonyms do not exist above the rank of genus.]

Available name A name published under the International Code of Zoological Nomenclature with a status equivalent to that of a validly published name under the International Code of Nomenclature for algae, fungi, and plants.

Avowed substitute See **replacement name.**

Basionym The legitimate, previously published name on which a new combination or name at new rank is based. The basionym provides the final epithet, name, or stem of the new combination or name at new rank (see also **name at new rank, new combination**).

Bibliographic error Mistake in the number of volume, or page, or other details in citation.

Binary combination (binomial) A generic name combined with a specific epithet to form a species name.

Binary name See **binary combination**.

Binding decision A recommendation made by the General Committee and ratified by an International Botanical Congress on (1) whether or not a name is validly published or (2) whether or not names are to be treated as homonyms. Binding decisions are listed in (1) App. VII or (2) VIII.

Binomial See binary combination.

Circumscription The diagnostic limit separating the individuals that balong to a taxon from all other individuals.

Citation The method of writing the original author's name after the binomial is called citation.

Combinatio nova (comb. nov.) See **new combination.**

Combination A name of a taxon below the rank of genus, consisting of the name of a genus combined with one or two epithets.

Compound A name or epithet that combines elements derived from two or more Greek or Latin words, a regular compound being one in which a noun or adjective in a non-final position appears as a modified stem.

Confusingly similar names Orthographically similar names at the rank of genus or below that are likely to be confused and are to be treated as homonyms if heterotypic or as orthographical variants if homotypic. Binding decisions may be made on whether or not the former are to be treated as homonyms.

Conserved name (nomen conservandum) (1) A name of a family, genus, or species, or in certain cases a name of a subdivision of a genus or of an infraspecific taxon, ruled as legitimate and with precedence over other specified names even though it may have been illegitimate when published or lack priority. (2) A name for which the type, orthography, or gender has been fixed by the conservation process.

Correct name The name that must be adopted in accordance with the rules for a taxon with a particular circumscription, position, and rank.

Cultivar The basic independent category used for organisms in agriculture, forestry, and horticulture and defined and regulated in the International Code of Nomenclature for Cultivated Plants.

Date of name The date of valid publication of a name.

Descriptio generico-specifica A single description simultaneously validating the names of a genus and its single species.

Description A published statement of a feature or features of a taxon; a description (or a diagnosis) is required for valid publication of a name.

Descriptive name A name of a taxon above the rank of family not formed from a generic name.

Designation The term used for what appears to be a name but that (1) has not been validly published and hence is not a name in the sense of the *Code* or (2) is not to be regarded as a name.

Diagnosis A statement of that which in the opinion of its author distinguishes a taxon from other taxa : a diagnosis (or a description) is required for valid publication of a name.

Diagnostic character Character or feature of a taxon that distinguishes from others.

Duplicate. Part of a single gathering of a single species or infraspecific taxon made by the same collector(s) at one time.

Effective publication Publication in accordance with the rule of nomenclature.

Element (as applied to typification). Applied to a specimen or illustration eligible as a type; also applied to a species name considered as the full equivalent of its type for the purposes of designation or citation of the type of a name of a genus or subdivision of a genus.

Epithet Used for the words in a combination other than

the generic name and any rank-denoting term; hyphenated words are equivalent to a single word. or any word in a name of more than one word, subsequent to the generic name.

Epitype A specimen or illustration selected to serve as an interpretative type when the holotype, lectotype, or previously designated neotype, or all original material associated with a validly published name, cannot be identified for the purpose of the precise application of the name to a taxon.

Et And.

Et aliorum, (et alii, et. Al.,) And of others.

Ex From or according to.

Exsiccata Dried and preserved specimens of plants.

Ex-type (ex typo), ex-holotype (ex holotypo), ex-isotype (ex isotypo), etc A living isolate obtained from the type of a name when this is a culture permanently preserved in a metabolically inactive state.

Filius Son of.

Final epithet The last epithet in sequence in any particular combination, whether in the rank of a subdivision of a genus, or of a species, or of an infraspecific taxon.

Forma specialis See **special form.**

Fossil-taxon A taxon (diatom taxa excepted) the name of which is based on a fossil type.

Gathering Used for a collection of one or more specimens made by the same collector(s) at the one place and time.

Heterotypic synonym (taxonomic synonym) A name based on a type different from that of another name referring to the same taxon; termed a "subjective synonym" in the International Code of Zoological Nomenclature and the International Code of Nomenclature of Bacteria (Bacteriological Code).

Holotype The one specimen or illustration used by the author, or designated by the author as the nomenclatural type.

Homonym A name spelled exactly like another name published for a taxon of the same rank based on a different type. Note: names of subdivisions of the same genus or of infraspecific taxa within the same species that are based on different types and have the same final epithet are homonyms even if they differ in rank, the rank-denoting term not being part of the name.

Homotypic synonym (nomenclatural synonym) A name based on the same type as that of another name termed an "objective synonym" in the International Code of Zoological Nomenclature and the International Code of Nomenclature of Bacteria (Bacteriological Code).

Hybrid formula An expression consisting of the names of the parent taxa of a hybrid with a multiplication sign placed between them.

Illegitimate name A validly published name that is not in accordance with specified rules of the nomenclature.

Illustration A work of art or a photograph depicting a feature or features of an organism, e.g. a picture of a herbarium specimen or a scanning electron micrograph.

Improper Latin termination A termination of a name or epithet not in accordance with the termination mandated by the Code.

Indelible autograph Handwritten material reproduced by some mechanical or graphic process (such as lithography, offset, or metallic etching).

Indelible Of written or printed matter that can not be rubbed out.

Indirect reference A clear (if cryptic) indication, by an author citation or in some other way, that a previously and effectively published description or diagnosis applies or that a basionym or replaced synonym exists.

Informal usage Usage of the same rank-denoting term at more than one non-successive position in the taxonomic sequence. Note: names involved in such usage are validly published but unranked.

Infrageneric Any taxon of a rank below that of a genus.

Infraspecific Any taxon of a rank below that of a species.

Invalid name A name which is not validly published.

Isoepitype A duplicate specimen of the epitype.

Isolectotype A duplicate specimen of the lectotype.

Isoneotype A duplicate specimen of the neotype.

Isonym The same name based on the same type, published independently at different times perhaps by different authors. Note: only the earliest isonym has nomenclatural status.

Isosyntype A duplicate of a syntype.

Isotype A duplicate specimen of the holotype.

Lectotype A specimen or illustration designated from the original material as the nomenclatural type if no holotype was indicated at the time of publication, or if the holotype is missing, or if a type is found to belong to more than one taxon.

Legitimate name A validly published name that is in accordance with the rules, i.e. one that is not illegitimate. (see also illegitimate name).

Manuscript Handwritten or typed material existing in only one or few copies.

Microfilm A copy of a published work or manuscript usually much reduced, on a continuous strip of transparent photographic material.

Misplaced term A rank-denoting term used contrary to the relative order specified in the Code.

Monotypic genus A genus for which a single binomial is validly published (see also unispecific).

Monstrosity A plant or specimen showing an abnormal structural condition.

Name A name that has been validly published, whether it is legitimate or illegitimate (see also designation).

Name at new rank (status novus) A new name based on a legitimate, previously published name at a different rank, which is its basionym and which provides the final epithet, name, or stem of the name at new rank. (see also **basionym, new combination**).

Name of a new taxon A name validly published in its own right, i.e. one not based on a previously validly published name; it is not a new combination, a name at new rank (status novus), or a replacement name (avowed substitute, nomen novum).

Nec Not of, nor of.

Neotype A specimen or illustration selected to serve as nomenclatural type if no original material is extant or as long as it is missing.

New combination (combinatio nova) A new name of rank lower than genus based on a legitimate,previously published name, which is its basionym and which provides the final epithet of the new combination (see also basionym, name at new rank).

New name A name as it appears in the place of its valid publication (see also nomenclatural novelty).

Nomen Name.

Nomen aliarum, (nom. al.) Names used by other authors.

Nomen alternativum. (nom. alt.) See **alternative family name.**

Nomen ambiguum, (nom. ambig.) Names used in different senses.

Nomen confusum, (nom. conf.) Name causing confusion.

Nomen conservandum (nom. cons.) See **conserved name.**

Nomen dubium, (Nom. dub.) Doubtful name.

Nomen novum (nom. nov.) See **replacement name.**

Nomen nudum (nom. nud.) A designation of a new taxon published without a description or diagnosis or reference to a description or diagnosis.

Nomen rejiciendum (nom. rej.) A name rejected in favour of a name conserved or a name ruled as rejected (see also **rejected name**).

Nomen utique rejiciendum (suppressed name) A name ruled as rejected under code. Note: it and all names for which it is a basionym are not to be used.

Nomenclatural novelty Any or all of the categories: name of a new taxon, new combination, name at new rank, and replacement name(see also **new name**).

Nomenclatural synonym See **homotypic synonym**.

Nomenclatural type The element to which the name of a taxon is permanently attached.

Non-fossil taxon A taxon the name of which is based on a non-fossil type.

Non Not of; used in the citation of later homonyms to indicate that these refer to taxa other than those described by the authors of the earlier homonyms.

Nothogenus A hybrid genus.

Nothomorph A term formerly denoting the only infraspecific rank, equivalent to variety, permitted within nothospecies. Names published as nothomorphs are now treated as names of varieties.

Nothospecies A hybrid species.

Nothotaxon A hybrid taxon.

Objective synonym See **homotypic synonym.**

Operautiqueoppressam See **suppressed works.**

Organism As used in this Code, the term is applied only to organisms traditionally studied by botanists, mycologists, and phycologists.

Original material The set of specimens and illustrations from which a lectotype may be chosen.

Original spelling The spelling employed when a name of a new taxon or a replacement name was validly published.

Orthography Spelling.

Orthographic error An unintentional mis-spelling.

Orthographical variants Various spelling, compounding, and inflectional forms of a name or its final epithet, only one nomenclatural type being involved.

Page reference Citation of the page or pages on which the basionym or replaced synonym was validly published or on which the protologue appears.

Paratype Any specimen cited in the protologue that is neither the holotype nor an isotype, nor one of the syntypes if in the protologue two or more specimens were simultaneously designated as types.

Parentheses Brackets.

Position Used to denote the placement of a taxon relative to other taxa in a classification, regardless of rank.

Priority A right to precedence established by the date of valid publication of a legitimate name or of an earlier homonym, or by the date of designation of a type.

Protologue Everything associated with a name at its valid publication, e.g. description, diagnosis, illustrations, references, synonymy, geographical data, citation of specimens, discussion, and comments.

Provisional name A designation proposed in anticipation of the future acceptance of the taxon concerned, or of a particular circumscription, position, or rank of the taxon.

Rank The relative position of a taxon in the taxonomic hierarchy.

Rejected name A name ruled as not to be used, either by formal action under code or overriding other provisions of the Code (see **nomen rejiciendum, nomen utique rejiciendum**) or because it was nomenclaturally superfluous when published or a later homonym.

Replaced synonym The legitimate or illegitimate, previously published name on which a replacement name (avowed substitute, nomen novum) is based. The replaced synonym, when legitimate, does not provide the final epithet, name, or stem of the replacement name.

Replacement name (**avowed substitute**, **nomen novum**) A new name based on a legitimate or illegitimate, previously published name, which is its replaced synonym and which, when legitimate, does not provide the final epithet, name, or stem of the replacement name.

Sanctioned name The name of a fungus treated as if conserved against earlier homonyms and competing synonyms, through acceptance in a sanctioning work.

Special form (forma specialis) A taxon of parasites, especially fungi, characterized from a physiological standpoint but scarcely or not at all from a morphological standpoint, the nomenclature of which is not governed by this Code.

Specimen A gathering, or part of a gathering, of a single species or infraspecific taxon made at one time, disregarding admixtures, mounted either as a single preparation or as more than one preparation with the parts clearly labelled as being part of the same specimen.

Status (1) Nomenclatural standing with regard to effective publication, valid publication, legitimacy, and correctness. (2) Rank of a taxon within the taxonomic hierarchy (see **name at new rank).**

Status novus (stat. nov.) See **name at new rank.**

Subdivision of a family Any taxon of a rank between family and genus.

Subdivision of a genus Any taxon of a rank between genus and species.

Subjective synonym See **heterotypic synonym.**

Superfluous name A name that, when published, was applied to a taxon that, as circumscribed by its author, definitely included the type of a name that ought to have been adopted, or of which the epithet ought to have been adopted, under the rules.

Suppressed name See **nomen utique rejiciendum.**

Suppressed works (opera utique oppressa) Works, ruled as suppressed, in which names in specified ranks are not validly published.

Synonym One of two or more names that apply to the same taxon (see **heterotypic synonym, homotypic synonym**).

Syntype Any specimen cited in the protologue when there is no holotype, or any of two or more specimens simultaneously designated in the protologue as types.

Tautonym A binary designation in which the specific epithet exactly repeats the generic name.

Taxon (taxa) A taxonomic group at any rank.

Taxonomic synonym See **heterotypic synonym.**

Teleomorph A meiotic sexual morph in pleomorphic fungi.

Type See **nomenclatural type.**

Type designation An explicit statement that establishes

the type of a name; either (1) a holotype or syntype(s) designated in the protologue or (2) a lectotype, neotype, or epitype subsequently designated under the provisions of the code and in accordance with it.

Typographic error An error in a printed text for which the printer is responsible not the author.

Unispecific With a single species.

Validate To make validly published; used in the context of a description or diagnosis, or illustration, effecting valid publication of a name.

Validlypublished Effectively published and in accordance with rule of the Code.

REFERENCES

1. Arthur, J.C. *et al.* (1904). Code of Botanical Nomenclature. *Bull.Torrey bot. Club,* 31:249–291.
2. Arthur, J.C. *et al.* (1907). American Code of Botanical Nomenclature. *Bull. Torrey bot.Club,* 34:167–291.
3. Bailey, L.H. (1949). Problems in Taxonomy. *Amer.bot.* 36:8–10.
4. Bartlett, H.H. (1940). The concept of genus. *Bull.Torrey bot. Club,* 67:349–362.
5. Bennet, S.S.R. (1979). An Introduction to Plant Nomenclature. Delhi.
6. Bhandari, M.M. (1978). Flora of the Indian Desert. Sc. Publ., Jodhpur.
7. Briquet, J. and A.B. Rendle. (1935). International Rules of Botanical Nomenclature-rev.Interna.botani.congr.Cambridge, 1930. Jena.
8. Brummitt, R.K. (1980). Questions of effective publication. *Taxon.* 29:477–484.
9. Cain, A.J. (1959). Deductive and inductive methods in post-Linnaean taxonomy. *Proc. Linn. Soc. Lond.* 170:185–217.
10. Camp, W.H., H.W. Rickett and C.A. Weatherby. (1947). International rules of Botanical Nomenclature. *Brittonia,* 6:1–120.
11. Camp, W.H. *et al.* (1947). International rules of Botanical Nomenclature. Supplement. Species lectotypicae generum Linnaei. *Brittonia* ,6:114–118.

12. Caspard Bauhin (1560–1624), Bergmann M, Wendler, D. Gegenbaurs Morphol. Jahrb. (1986); 132 (2):173–81.

13. Clapham, A.R.,Tutin, T.G. and Warburg,E.F. (1962). Flora of the British Isles, 2nd edition, Cambridge Univ. Press.

14. Cronquist, A. (1968). The Evolution and Classification of Flowering Plants. Houghton Mifflin, Boston.

15. Dahlgren, R.M.T and Ib. Friis. (1981). A note on effective publication. *Taxon*. 30:499–501.

16. Davis, P.H. and Heywood, V.H. (1963). *Principles of Angiosperm Taxonomy*. Edinburgh and London.

17. DeCandolle, A.P. (1813). *Theorie elementaire de la botanique*, Paris

18. DeCandolle, A.P. (1824-73). *Prodromus systematic naturalis* regni vegetabilis. Paris, 17 vols.

19. De Wit, H.C.D. (1963). Plants of the world: The Higher Plants. 2:1–340.Thames and Hudson, London.

20. Dobzhansky, T. (1937). Genetic nature of species differences. *Amer. Nat.* 74:404–420.

21. Du Rietz, B. (1930). The Fundamental basis of biological taxonomy. *Svensk Bot. Tidskr.* 24:333–428.

22. Eichler, H.J. (1981). A further note on the effective publication. *Taxon.* 27:128–129.

23. Fosberg, F.R. and Sachet, M.H. (1965). Manual for Tropical Herbaria. *Reg. Veg.* Vol. 39, Utrecht – Netherlands.

24. Green, M.L. (1927). History of Plant Nomenclature, 403–415, London.

25. Greuter, W. and Voss E.G. (1982). Report on Botanical Nomenclature [Sydney 1981]. *Englera.* 2:1–124.

26. Greuter, W. and J. McNeill. (1987). Synopsis of proposals on botanical Nomenclature [Berlin 1987]. *Taxon.* 36:174–259.

27. Greuter, W. *et al.* (1988). International Code of Botanical Nomenclature, *Reg. Veg.* 118, Berlin.

28. Hawksworth, D.L. (1995). Steps along the road to a harmonized bionomenclature. *Taxon.* 44:447–456.

29. Heller, J.C. (1964). The early history of binomial nomenclature. *Huntia.* 1:33–70.

30. Henry, A.N. and Chandrabose, M. (1980). An aid to the International Code of Botanical Nomenclature. Today and Tomorrows Printers and Publishers, New Delhi.

31. Heywood, V.H. (1963). Principles of Angiosperm Taxonomy. Academic Press, London and New York.

32. Heywood, V.H. (1974). Systematics of the Stone of Sisyphus. *Biol. J.Linn.Soc.* 6:169–178.

33. Heywood, V.H. (1976). *Plant Taxonomy.* 2nd edn. London.

34. Hitchcock, A.S. (1919). Type-basis Code of Nomenclature of Taxonomy. *Science.* 49:333–336.

35. Jain, S.K. and Rao, R.R. (1977). *A Handbook of Field and Herbarium Methods.* New Delhi.

36. Jeffrey, C. (1968). *An Introduction to Plant Taxonomy.* J and A Churchill Ltd. London.

37. Jeffrey, C. (1977). *Biological Nomenclature.* 2nd edn. Great Britain.

38. John Ray. 1686–1688. Historia Plantarum, 2 vols. London.

39. John, H.St. (1958). *Nomenclature of Plants.* New York,.

40. Jussieu, A.L.de. (1789). Genera plantarum. Paris.

41. Lanjouw, J. *et al.* (1950). Botanical Nomenclature and Taxonomy- a symposium at Utrecht. (1948). Netherlands.

42. Lanjouw, J. *et al.* (1952). International Code of Botanical Nomenclature, Stockholm (1950), Netherlands.

43. Lanjouw, J. (1966). International Code of Botanical Nomenclature, Edinburgh 1964, Netherlands.

44. Lawrence, G.H.M. (1951). *Taxonomy of Vascular Plants.* 823pp. Macmillan, New York.

45. Lawrence, G.H.M. (1969). *Taxonomy of Vascular Plants.* (Reprinted.) Oxford and IBH, Calcutta.

46. Linnaeus, C. (1753). *Species Plantarum*, Stockholm, 2 vols.

47. Mayr. E. (1942). Systematics and the origin of species. Columbia University Press, New York.

48. McNeill, J. (1981). Report of the committee on generic typification. *Taxon*. 30:200–207.

49. McNeill, J. (1981). Typification of generic names *Ibid*. 457–462.

50. McVaugh, R. *et al.* (1968). An annotated glossary of Botanical Nomenclature. *Reg. Veg.* Vol. 56. Utrecht- Netherlands.

51. Meikle, R.D. and Brummitt R.K. (1981). A further proposal on typification of generic names. *Taxon.* 30:501–504.

52. Nicolson, D.H. (1980 a). Key to the identification of effectively/ ineffectively published material. *Taxon.* 29:485–488.

53. Raizada, M.B. (1968). What is there in a name? Why change? *Indian Forester.* 94:37–46.

54. Rickett, H.W. (1941). Linnaean Rules of Nomenclature. *Torreya* 41:188–191.

55. Rickett, H.W. and W. H. Camp. (1950). The application and use of botanical names. *Bull. Torrey bot. Club,* 77:245–261.

56. Santapau, H. (1964–65). The International Code of Botanical Nomenclature. 51st. and 52nd *Session of Indian Science Congress-Presidential Address*, Calcutta.

57. Santapau, H. (1965). The International Code of Botanical Nomenclature. *Science and Culture*. 31:456–467.

58. Santapau, H. and Henry, A.N. (1973). *A Dictionary of the Flowering Plants in India*. New Delhi,.

59. Shukla, P. and S.P. Misra. (1982). *An Introduction to the Taxonomy of Angiosperms.* Vikas Publ. House. India.

60. Sivarajan, V.V. (1981a). Taxonomic notes on genus *Portulaca* Linn. *Jour. Bombay Nat. Hist. Soc.* 78:256–260.

61. Sivarajan, V.V. (1984). *Introduction to principles of plant Taxonomy*. Oxford and IBH Publ. Co. New Delhi.

62. Smith, A.C. (1957). Fifty years of Botanical Nomenclature. *Brittonia* 9:2–8.

63. Sprague, T.A. (1924). Proposed changes in the International Rules, *Jour. Bot.* 62:196–198.

64. Stace, C.A. (1980). *Plant Taxonomy and Biosystematics*. Edward Arnold, London.

65. Stafleu, F.A. *et al.* (1972). *International Code of Botanical Nomenclature*. Seattle 1969. *Reg. Veg.* Vol. 97 Utrecht-Netherlands.

66. Stafleu, F.A. *et al.* (1978). *International Code of Botanical Nomenclature*. Leningrad, 1975. *Reg. Veg.* Vol. 97. Utrecht-Netherlands.

67. Stearn, W.T. (1966). *Botanical Latin*. Great Britain.

68. Steudel, E. G. 1840–41. Nomenclature botanicus.

69. Tournefort, J.P.DE. (1719). Institutiones Rei Herbariae, ed. 3. 1, 508. Paris.

70. Weatherby, C.A. (1942). Subspecies. *Rhodora*. 44:157–167.

71. Weatherby, C.A. (1946). Changes in Botanical names. *Amer. Midl.Nat.* 35:795–796.

72. Weatherby, C.A. (1949). Translation of remarks of A. De Candolle. *Amer. Jour. Bot.* 36:7.

73. Fiori, A. and G.Paoletti. (1896). Flora analitica d'Italia 1:1–256. Padova.

74. Hanelt, P. and J. Schultze-Motel. (1983). Proposal (715) to conserve *Triticum aestivum* L. (1753) against *Triticum hybernum* L. (1753) (*Gramineae*). *Taxon*. 32:492–498.

75. Trehane, P. *et al.* (1995). International Code of Nomenclature for Cultivated Plants. *Regnum Veg.* 133.

76. Frizzell, D.L. (1933). Terminology of Types. *American Midl. Nat.* 14:631–668.

77. Linnaeus, C. (1737). Critica botanica. Leyden.

78. Linnaeus, C. (1737). Flora Lapponica. Amsterdam.

79. Linnaeus, C. (1737). Genera Plantarum. Lugduni Batavorum.

80. Linnaeus, C. (1737). Hortus Cliffortianus. Amsterdam.

81. Greuter, W., Mcneill, J. Barrie, F. R. Burdet, H.M. Demoulin, V. Filgueiras, T.S. Nicolson, D.H. Silva, P.C. Skog, J.E. Trehane, P. Turland, N.J. Hawksworth D.L. (editors & compilers). (2000). International Code of Botanical Nomenclature (St. Louis Code) adopted by the Sixteenth International Botanical Congress, St. Louis, Missouri, July–August. 1999. *Regnum Veg.* Vol.138.

82. Engler, A. (1892). Syllabus der pflanzen familien. Berlin.

83. Degener, O. (1940). Flora Hawaiiensis. Vol.4, Published by author.

84. Pliny. 32–75 A.D. "The natural history of Pliny" (1855). Pliny the Elder, Bostock, John in: Pliny vol 2:London. H.G.Bohn.

85. Plukenet, L. (1692). Phytographia tab. 240 fig.2. London.

INDEX

Symbols

A

C

D

H

I

N

Q

R

U

V

Printed in Dunstable, United Kingdom

91222489R00170